Beyond Mobile

£25.00 S2
15763

Beyond Mobile

People, communications and
marketing in a mobilized world

by

Mats Lindgren
Jörgen Jedbratt
Erika Svensson

palgrave

© Mats Lindgren, Jörgen Jedbratt and Erika Svensson 2002

All rights reserved. No reproduction, copy or transmission of this publication may be made without written permission.

No paragraph of this publication may be reproduced, copied or transmitted save with written permission or in accordance with the provisions of the Copyright, Designs and Patents Act 1988, or under the terms of any licence permitting limited copying issued by the Copyright Licensing Agency, 90 Tottenham Court Road, London W1T 4LP.

Any person who does any unauthorised act in relation to this publication may be liable to criminal prosecution and civil claims for damages.

The authors have asserted their right to be identified as the authors of this work in accordance with the Copyright, Designs and Patents Act 1988.

First published 2002 by
PALGRAVE
Houndmills, Basingstoke, Hampshire RG21 6XS and
175 Fifth Avenue, New York, N.Y. 10010
Companies and representatives throughout the world

PALGRAVE is the new global academic imprint of
St. Martin's Press LLC Scholarly and Reference Division and
Palgrave Publishers Ltd (formerly Macmillan Press Ltd).

ISBN 0–333–98508–7 hardcover

This book is printed on paper suitable for recycling and made from fully managed and sustained forest sources.

A catalogue record for this book is available from the British Library.

Library of Congress Cataloging-in-Publication Data

Lindgren, Mats, 1950–
  Beyond mobile : communications and marketing in a mobilized world / by Mats Lindgren, Jörgen Jedbratt, Erika Svensson.
    p. cm.
  Includes bibliographical references and index.
  ISBN 0–333–98508–7
  1. Cellular telephone systems—Marketing. 2. Mobile communication systems—Marketing. I. Jedbratt, Jörgen. II. Svensson, Erika. III Title.

HE9713 . L56 2002
384.5'35—dc21
                                          2001057742

Editing and origination by
Aardvark Editorial, Mendham, Suffolk

10 9 8 7 6 5 4 3 2 1
11 10 09 08 07 06 05 04 03 02

Printed and bound in Great Britain by
Creative Print & Design (Wales), Ebbw Vale

Contents

List of Figures and Tables	ix
Foreword	xii
Executive Summary	xiii
The greater framework	xiv
The smaller framework – the mobile marketplace	xiv
The mobile individual	xv
Four scenarios for tomorrow's mobile marketplace	xv
Acknowledgments	xix
Introduction	xxii
Background and purpose	xix
Method, implementation and readers' advice	xix

Part I   **The Context**   1

Chapter 1   **What is the Mobile Marketplace All About – Really?**   3
- The human perspective   3
- The perspective of the marketing company   4
- … and the media perspective   4
- Is the mobile marketplace a mobile Internet?   4
- Summary   7

Chapter 2   **The Topography of Mobility, Information and the Market**   8
- Movement   10
- Information handling   13
- Commerce   17
- The mobile marketplace   22
- Summary   23

Chapter 3   **The Individual as a Mobile Creature**   24
- Mobile archetypes, states of mind   24
- Mental mobility and games of identity   27
- Different kinds of mobile service   29
- The mobile purchasing cycle   34
- Summary   36

Contents

Part II		**Preconditions, Driving Forces and Trends**	**37**
Chapter 4		Technology, Institutions, People and the Economy	39
		Social change as a game of push and pull	41
		Technology and people – a necessary whole	43
		The S-curve and driving forces in various phases	45
		Summary	50
Chapter 5		**Technology**	**51**
		Moore's law – and Metcalfe's	52
		Network benefits are exponential	53
		Mobile telephone networks	54
		Wireless local networks	56
		Bluetooth – wireless micro-networks	57
		Positioning	58
		Technology today and in the future	59
		Technical trends	64
		Technological aspects of data transfer and "always-on" Internet connection	72
		Connected technology	73
		Technology for handling information	73
		Digression: What happened (and is happening) to the software agents?	74
		A prognosis?	77
		Summary	79
Chapter 6		**The Institutions**	**80**
		The importance of institutions	80
		The emergence of the knowledge-industrial society	81
		Rules of play, legislation and state incentives	88
		Summary	94
Chapter 7		**The Individual**	**95**
		Postmodern values – the victory of pluralism?	96
		What will we want to consume?	97
		The importance of the brand	97
		Open landscapes or the tyranny of diversity?	98
		Twenty consumer trends	100
		Summary	111
Chapter 8		**The Economy**	**112**
		Gigantic investments	112
		Where are the revenues?	115
		Two markets and two consumer groups	116
		Summary	119
Chapter 9		**The New Marketing Logic**	**120**
		The customer-driven marketplace	120
		The evolution towards micro-geographical marketing	125
		Position commerce	127

		Media trends	128
		Marketing trends	139
		Summary	146
Part III		**The Arena**	**147**
Chapter 10		The Players in the Mobile Marketplace	149
		A map of the players	149
		Companies, media and receivers become one	152
		Marketing and selling companies (and organizations)	153
		Advertising and media agencies, intermediaries	154
		Traditional media	155
		New interactive media	158
		Content providers	159
		Content packagers	160
		The portals	161
		Position players	162
		Mobile network and service operators	163
		Mobile retailers	167
		System and platform providers	167
		Software companies	168
		Mobile manufacturers	169
		The mobile individual	171
		Summary	174
		Digression: Experts and users give their views on tomorrow's mobile marketplace	176
		The Swedish population as users	176
		Conclusions	185
		Summary	188
		Experts pronounce on the mobile marketplace – a "Delphi study"	189
		Summary	194
Part IV		**Mobile Marketing for Tomorrow**	**197**
Chapter 11		The Mobile Marketplace in 2007 – A Future Free of Surprises	199
		A brief outline of the scenarios and scenario methodology	199
		Safe predictions	201
		Consequences for media agencies	204
		Consequences for marketing companies	205
		Scenario for the future – a user family in 2007	220
		Summary	223
Chapter 12		**Four Scenarios for a Mobile Tomorrow**	**224**
		Main uncertainties	224
		A joker in the pack	227
		Tomorrow's mobile marketplace – four scenarios	229
		Conclusion	236

Postscript	Mobile Tomorrows	243
	Kairos Future	244
Glossary		245
Notes		250
Sources		253
Index		257

List of Figures and Tables

Figures

0.1	A description of the book's structure and working model	xxi
1.1	The three dimensions of the mobile marketplace	6
1.2	The mobile marketplace merges the virtual and physical marketplaces into something qualitatively new	6
2.1	The significance of car ownership in the 1950s	12
2.2	Wisdom presupposes both high-level access to information and a high degree of reflection	17
2.3	The Model T Ford	20
3.1	The dimensions of mobility	26
3.2	The mobile marketplace: time- and position-critical services	30
3.3	The buying cycle	35
4.1	The interplay between technology, people and institutions	40
4.2	The introduction of new technology in a market often takes the form of an S-curve	46
4.3	Interplay between expectations and actual development	47
5.1	The cost of computation power calculated in MIPS	52
5.2	New technology breaks through at a quickening pace	53
5.3	Transfer capacity and range of various wireless networks	57
5.4	Mobile phone and Internet use in different groups of countries	59
5.5	Number of I-mode users	61
5.6	The emergence of a metanet	65
5.7	Minaturization for user-friendliness	66
5.8	Mobile devices of the future	68
5.9	Microsoft's Tablet PC	69
6.1	The price of a haircut in relation to shampoo over a 20-year period	83

List of Figures and Tables

8.1	Norwegian Telefonor's mobile service DJUICE targets youth in Scandinavia, New Zealand, Thailand and Singapore	117
8.2	The market for mobile services	119
9.1	The product-driven system	121
9.2	The distribution-driven system	122
9.3	The customer-driven system	123
9.4	The mobile as a membrane between the personal universe and the external world	124
9.5	Influencing purchasing decisions in the shop	126
9.6	The media day	134
9.7	Value formula for future media	136
9.8	*Metro*, distributed free in public transport hubs all over the world	137
9.9	Polarization between attractive concepts and exclusive content	138
10.1	The players involved in communication in the mobile marketplace	150
10.2	The merging of players in the mobile marketplace	152
10.3	The modular media map	157
10.4	The role of position players	163
10.5	The conglomerate Virgin – with its Virgin Mobile – was the first mobile operator without its own network	165
10.6	The world's biggest manufacturers of mobile systems in 1999	168
10.7	The 21st-century knight in armor is considerably more lightly dressed than his predecessor	173
10.8	Summary of areas of interest, resources and possible strategies of various players in the mobile marketplace	175
10.9	Survey: Preferences for various services of the next generation of mobile telephony	180
10.10	Two principal groups in the mobile marketplace: Moklofs and Yupplots	182
10.11	The view of experts: mobile services which will have established themselves (reached 30 percent saturation) by the year 2005	194
10.12	Experts' assumptions about development up to 2015	195
11.1	Scenario building	201
11.2	Choosing the customer as the strategy	207
11.3	Mental states of consumers	211
12.1	Overview of the four scenarios	230

Tables

1.1	Examples of different types of personalization	5
2.1	Work, education and entertainment – change over four epochs of human history	9

List of Figures and Tables

2.2	The development of the markets from barter to an excess market	21
3.1	Dimensions of mobility	25
3.2	Various mobile services and willingness to pay	31
4.1	For new technology to break through, four subsystems have to be in place: technology that works, systems for data and information handling, infrastructure and functioning content	45
4.2	A comparison between the motivations of different user groups	48
5.1	Highest theoretical data transfer speeds of various mobile data communications. In practice, speeds will be considerably lower	55
5.2	Mobile customers year 2010 according to NTT DoCoMo	63
8.1	License methods, the number and costs	114
10.1	Comparison between potential mobile portal players	161
10.2	Telephone operators year 2000	164
10.3	The world's largest manufacturers of mobiles, market share Jan–March 2001	170
10.4	Three types of mobile services. Multivariate analysis shows that the services fall into three distinct groups	185
10.5	Interest for services among the three early adopter groups Moklofs, Yupplots and Sallies	186
11.1	Examples of differences between prognoses, scenarios and visions	200
11.2	Summary of general assumptions of the coming years	204
11.3	Traditional media and the mobile marketplace	215
12.1	Examples of the effects of various radio transmitters	228
12.2	Scenario grids	237

Foreword

We welcome this book.

One of the great things about the future – a future "beyond mobile" – is that there are no rules. The mobile marketplace is still very much in its formative years, a market characterized by hostile mergers and acquisitions, and fierce competition, not only for network operators, but handset manufacturers, and content providers. Thus the debate governing the future of this market is complex, ever changing, and here to stay.

The advent of multichannel access through the convergence of distribution pipes means that consumers are now able to access information and their favorite brands from a variety of devices: phone, PC, PDA and even a games console through wireless LANs, Bluetooth broadband and UMTS networks. The challenge to companies though, is to harness this technology for the benefit of their customers, since understanding and delivering on their needs and motivations will ultimately win the war for long-term, high-value customers.

And it is those companies that have embraced the changing context and driving forces placed on the mobile marketplace over the last ten to twenty years (most notably, the movement from product-driven to customer-driven business practices) that now find themselves best placed to deliver genuine customer benefits, and thus win this war.

Given the emergence of the customer champion, with the word "product" having been replaced by the mantra "customer," this heralds a new phase for companies. The power no longer resides with the brand owner, but rather the end customer. It is now customers who drive NPD, CRM and other capabilities, through the demands that they place on these companies. Coupled to this, and given the advent of and hype surrounding relationship marketing in preference to mass media communications, how will customers best utilize this avenue to get their messages across in an increasingly complex and crowded mobile marketplace? The theory suggests that targeted one-to-one marketing is one of the most creative and rewarding ways of communicating with your (current and potential) customers. In practice, however, the full potential of this relatively new avenue has yet to be realized.

The challenge becomes further magnified in the global mobile marketplace given the changing demands of different countries. The need for global consistency across naïve to relatively mature markets dictates that business models should be flexible enough to manage both acquisition- and retention-led propositions. The challenges in the Finnish market are clearly very different to those in Romania.

Interestingly, many companies are well positioned to provide national, even regional (European and North American), mobile services. To date, however, no company, nor for that matter any alliance or joint venture, has been able to demonstrate enough credibility to dominate the global arena. A handful of companies such as AT&T, NTT DoCoMo, Vodafone and Orange is actively seeking to create this global footprint, but as yet little consistency between markets, products and service levels prevails – largely because of corporate heritage, government regulations and cultural differences. Each company has a rigorous and considered business strategy to take it forward, but it is the brand that will provide the desired levels of global consistency, loyalty, and wealth creation.

If one looks at the latest Interbrand publication in *Business Week* (August 2001), it is interesting to note how AT&T and Nokia both secured a top ten ranking in the Interbrand top 100 Billion Dollar brand league table. Furthermore, with only one network provider and one handset manufacturer featured in the top 25, this would suggest that more consolidation and shakeout is still to come. Particularly since, as we move further into the future, we will see the competitive mobile landscape becoming increasingly complex, as the focus shifts from network to service and content provision, with new competitors such as AOL, IBM, BBC and MTV entering the fray.

Great brands create wealth and provide the brand owner with a security of earnings they would otherwise not enjoy. At a time when brand value guides shareholder value, we need to protect what is perhaps our most valuable asset, our brand. And this is as important in mobile telephony as in any classically branded company or industry. By way of some perspective on that value, let us look at the UK. In August 1999, One2One was sold to Deutsche Telecom for £6.9bn, equivalent to £2,460 per customer. Two months later Orange was sold to Mannesmann for £19.8bn, equivalent to £5,400 per customer – 120% more than One2One. Orange was later sold to France Telecom for a cumulative value of £31bn. At that time, both One2One and Orange had similar subscriber numbers, network and operational capabilities but Orange, with arguably the strongest brand in its field, and hence the potential for greater customer lifetime value, was clearly the more valuable and valued business.

The future of the mobile marketplace is ever changing and *Beyond Mobile* brings that future closer to all of us with many interesting insights and anecdotes.

RICHARD WARD and TOM BLACKETT
Interbrand Group

Executive Summary

We are all of us being drawn into the electronic world, and we can't stop it. It's like being given a car without anyone telling you how to drive it, and you don't have a road map. We're driving blind. (Sandy Sparks, Lawrence Livermore Laboratories)[1]

None of us knows anything for certain about the future. There are no detailed maps. Nonetheless, this books sets out to look at the mountains, valleys, tropical beaches and deserts we might expect to find on a map of the mobile marketplace.

The modern marketplace is a broad phenomenon. It extends across technologies, organizations, medias and people, and it encompasses everything from human behavior to investment planning.

In order to navigate among all these divergent perspectives we have drawn up certain dividing lines, focusing on particular areas and viewing the modern marketplace from two broad perspectives.

The first of these is what might be thought of as "the human factor" – that is, the ways in which people may be expected to utilize the new technological possibilities of constant Internet connectivity.

And second, how companies engaged in marketing or selling in this environment can use the possibilities of the mobile marketplace.

We have chosen to place these perspectives within two larger contexts, which, for the sake of simplicity, may be referred to as a "smaller" or "greater" framework.

The smaller framework is the area formed by many different players, from system providers like Ericsson, Lucent, Nokia and others, to newly established mobile service providers.

The greater framework is the society and time in which the mobile marketplace is evolving – what for the sake of directness might be called a society in transition from industrial to knowledge-industrial. The greater framework is thus characterized by increasing globalization, technocracy, postmodern values and a national economy based on knowledge and experience.

Let us briefly summarize the most important findings.

The greater framework

There is nothing coincidental about the fact that the mobile marketplace is emerging at this precise point in time. The main point, of course, is that the technical prerequisites for the development of a mobile marketplace have only recently been put in place. With increasing competition between companies and customers, increasing numbers of niche markets, and increasingly narrow sectors, there is an obvious need for a continuous, personalized dialogue with customers. The mobile marketplace fits the bill exactly – provided that customer data can be collected and sorted, and communications tailor-made in cost-effective ways. With the emergence of knowledge organizations and knowledge employees out there in the field, in constant communication with their companies, there is a clear need for companies to develop mobile solutions.

Last but not least are the human possibilities. A large proportion of the population in the West already has mobile phones and the Internet. Fundamentally we seem to be social creatures that like to move around. However, wherever we are, we want to keep in touch with the rest of the group. The global youth culture that has never known a world without phones, the Internet and video games, would seem to be a natural target group for selective mobile services.

Obviously, there are also many question marks: is the technology going to be good enough, quickly enough? Will the revenue streams generated be proportionate to the gigantic investments required to develop third- (and fourth-) generation mobile phones? Do we really want to leave electronic footprints wherever we go?

The smaller framework – the mobile marketplace

A host of different players in the mobile marketplace are already jockeying for position in the future – everything from infrastructure companies, content providers, packaging companies and consumers. It is a complex picture. The roles are not yet clear. Many of the players have more than one role at the same time. Content providers with strong brand names are using their position to make themselves independent from the distributors.

Mobile operators are doing everything in their power not to "miss the train again" as they did in the early days of the Internet. They need to do as much as they can to recoup their investments in licenses and networks – and to do this, they are investing in portals and content. Manufacturers of mobile phones and palmtop devices are fighting for dominance in the future market. Probably they will also encounter competition from own-brand companies and even high street retailers.

The mobile individual

What types of services would we like to have access to? A great deal of evidence points to two types – services in the entertainment sphere, and services that increase personal productivity. If we take a look at the groups already using mobile services, the largest of them is tied to entertainment services such as network-enabled games, chat and other community functions. An example of this is Moklofs – Mobile Kids with Lots of Friends. In terms of the productivity-based services, demand is highest among Yupplots – Young Urban Professional Parents with Lack of Time. But there are many unknown factors. How accessible do we really want to be? How much information can we handle? When does "broadband overload" kick in?

Four scenarios for tomorrow's mobile marketplace

Looking a few years ahead, the decisive question is the speed with which the mobile marketplace will break through. In truth, speed has to do with a range of factors such as simplicity and utility. In order to achieve widespread usage, the services have to be easy to use and useful. They must also have broad appeal. There has to be a readiness among consumers to be constantly connected, in return for perceived monetary rewards. Suggesting that all this will be in place within five to seven years is not a secure prediction.

Another uncertainty which particularly affects the mobile marketplace as an advertising channel, is about people's attitude to privacy. Will we be prepared to leave an electronic footprint, so that tailor-made information can be returned to us? Will we do so to such an extent that the mobile becomes a significant advertising channel? All these factors have ramifications not only for the financing of the networks, but also in relation to what sorts of players will be able to achieve success in the mobile marketplace.

Using these two uncertainties as a starting point, we see four major scenarios up to the year 2007.

Scenario 1: Mobile Klondike 2007

In the Mobile Klondike, developments have exceeded the wildest expectations. The vast majority of the population in the West not only have mobiles, but regularly use mobile services. The mobile Internet is much bigger than the traditional Internet.

Positional services have become a killer application and few people express any trepidation about handing out information to various suppliers, or allowing themselves to be positioned. The mobile Internet is technically easy to use, hence service providers have not had a difficult task establishing themselves in the mobile marketplace. There is a host of different portals and channels used by a wide variety of people.

The physical and mobile marketplaces are rapidly merging into an integrated marketplace. More and more people are talking about the "metanet" – that is, the integration of the physical and virtual environments.

Scenario 2: Trusted Guide 2007

In the Trusted Guide scenario, the development of the mobile marketplace has been hampered to some extent by a fierce privacy debate. This has meant that positional services are more or less a flop. There has also been a strong reaction against the holding of information about customers. However, some large, well-established brands have managed to build up trust to the extent that many customers are prepared to hand out their personal details, in order to be eligible for special offers. Telecommunications companies and supermarket chains are examples of the latter.

Scenario 3: Professional Users 2007

Professional Users is the most pessimistic scenario. Demand, services, prices, low numbers of phones and not least the privacy debate have all had a negative effect on the development of the mobile marketplace. The only successful development has been a mobile extension of corporate intranets. The primary users are professionals, perhaps sales people working in the field, or other knowledge-based workers who need to be constantly in touch with their company, clients and partners. No marketplace in the real sense has emerged, and weak demand has led to serious problems for the entire telecommunications sector.

Scenario 4: Community Lifestyle 2007

In the Community Lifestyle scenario, while a mass market has not yet emerged, more than purely professional users are nonetheless accessing the services. Many point to the fact that Alexander Graham Bell was actually right when concluding about his new patented phone: It's only a toy ... For this is precisely how the mobile phone is viewed by the majority: as a toy.

Those who have picked up on the new possibilities are above all children, teenagers and young adults between the ages of 10 and 30. They use the mobile phone as a communication and games platform. Nintendo and Sony were the first to bring out platforms for networked games, which also functioned as telephones, cameras, and so on. In terms of sales of mobiles, Sony and Nintendo have also overtaken most of the big mobile phone manufacturers of the 1990s.

Acknowledgments

This book is the public manifestation of a project running for several years to look into the future of the mobile marketplace. Much of the book is based on the findings of a six-month multiclient project on the mobile individual and the mobile marketplace as an advertising channel. We want to take this opportunity of thanking our project sponsors: MediaAcademy – a company in the MediaCom group, MediaCom, TV4, *Svenska Dagbladet*, Posten, Wezapp.

In the course of this journey, we have interviewed scores of people, listened to experts and gauged public opinion on mobile technology. We have discussed our findings and the content of this book, with participants and delegates in seminars, lectures and workshops. To all those who have contributed with views, suggestions, criticisms and encouragement, we wish to extend our thanks.

Crucial help has also been forthcoming from our colleagues at Kairos Future.

We particularly want to thank Per Florén for his contribution on software agents, Johanna Laurent, Linda Tjernberg, Mikael Raaterova, Christian Lernberg, Per Snaprud and Jennie Sjögren for their help in areas such as research, analysis, drafting of notes and reports, ideas, and so on.

MATS LINDGREN, JÖRGEN JEDBRATT
AND ERIKA SVENSSON

Every effort has been made to trace all the copyright holders but if any have been inadvertently overlooked the publishers will be pleased to make the necessary arrangements at the first opportunity.

Introduction

Background and purpose

The aim of this book was to deepen our understanding of the possibilities of the mobile marketplace, and to do so from a nontechnological perspective. Rather than focusing on technology, we have chosen to direct most of our attention at "the mobile connected individual" and the mobile marketplace as an advertising channel.

The mobile marketplace is an uncertain area to move into. Do we really want a mobile marketplace? If so, what do we actually want from it? Will it be full of advertising and, if so, what will be the interplay between mobile, interactive advertising and traditional advertising? And last but not least: Where is there real money to be made?

Because the level of insecurity is high, we chose to develop this project as a scenario study. We have attempted to sort certain from uncertain propositions, and around the strategic uncertainties we have built a number of scenarios, of which four are presented in this book.

Method, implementation and readers' advice

The majority of this examination of tomorrow's mobile marketplace has been carried out by an internal group of consultants at Kairos Futures, and principally by the authors of this book.

Working with participants in the project, a series of seminars were held where significant factors and trends for the future of the mobile marketplace were identified. In addition, three special studies were undertaken. The first of these was a series of deep interviews with prominent experts[2] – in areas such as technology and consumer affairs – on issues expected to be important for the mobile marketplace. The second special study had broader scope. Professionals in the mobile Internet and "the New Economy" were

Introduction

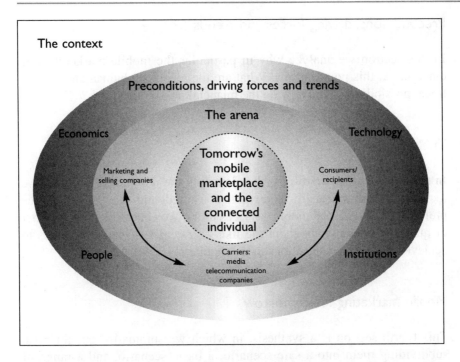

Figure 0.1 A description of the book's structure and working model

asked to evaluate future developments in the mobile marketplace. The third special study was a questionnaire, in which 2000 Swedish users of mobiles were asked how they were interested in employing mobile technology.

The main body of the book is divided into four parts, which largely reflect our research method. The book can either be read chronologically, or alternatively, used as a reference source. Figure 0.1 describes the book's structure and working model.

The context

The first section of the book attempts to place the mobile marketplace and the mobile individual in a larger context and longer term perspective.

Preconditions, driving forces and trends

In this section we analyze why in particular the mobile marketplace is emerging at this point in time. What are the technical, human and institutional possibilities? And how do the financial possibilities look?

The arena

In the third section, we discuss the players in the mobile marketplace. We recount the assessments of the experts consulted, and also summarize the preferences of potential mobile services users. We conclude with a number of qualified assumptions about the likely strategies of various players to achieve advantageous positions in the mobile marketplace.

Mobile marketing for tomorrow

This fourth section is a synthesis, in which we summarize our findings, subdividing them into a safe scenario, a basic scenario, and a range of strategic uncertainties developed in a number of other scenarios. We discuss the consequences for players in the market, both of the safe scenario and some of the other alternatives. Readers might find that it is to their advantage to read this section first, before studying earlier sections in the book.

In many places we use abbreviations of technical or other terms. However, we wish to point out that there is a glossary at the back of the book that clarifies most of the terminology.

PART I

The Context

What are the implications of the mobile marketplace, when analyzed from a broader, more long-term perspective? Is the mobile marketplace a natural progression, or does it represent a qualitative leap – something entirely new?

These questions are discussed in this first part.

After a quick overview of the mobile marketplace, we go on to take a deeper look at the interplay between the Internet, mobile telephony and the mobile marketplace. We analyze the mobile marketplace using a more long-term perspective that includes the history of mobility, information and the media marketplace. Finally we discuss the individual as a mobile creature. What kinds of mobile services seem imminent within the next few years?

Chapter 1

What is the Mobile Marketplace All About — Really?

If one takes a long-term historical perspective, there are compelling reasons for seeing the mobile marketplace as the next step of a developmental process that has been underway for many hundreds of years – and that will finally culminate in individualization, the immediate satisfaction of needs and a longing for freedom.

The human perspective

The mobile marketplace is the perfect expression of the three-pronged revolution of the late 1900s: the revolution of possibility, of knowledge and of freedom.

From this perspective, the mobile marketplace is a natural development. It enables us to migrate in our behavioral patterns from the Internet and IRL-meetings (in real life) to a broader group of people in our social group and in the marketplace at large. It is a phenomenon that creates freedom; it gives us choices about *when* and *where* we carry out certain tasks. It liberates us from the limitations of physical space.

We base our discussion on the idea of the individual as a consumer in the marketplace. The consumer is a player – perhaps even a king? With the mobile marketplace never further away than an arm's length, the consumer will always have the option of bringing the mobile marketplace into contact with the physical marketplace. In the shop, he/she can choose between the mobile and the physical. Henceforth, there will always be a choice, wherever the consumer finds him/herself.

The perspective of the marketing company

For the marketing company, the mobile marketplace represents a threat as much as it does an opportunity. With a mobile marketplace available at the touch of a button, every consumer becomes King. As a physical player in the market, you will have customers that carry out comparison-shopping on your premises. They can try out your products, taste them and touch them, only to buy them from someone else.

But the mobile marketplace is also an opportunity for a company with a physical (as opposed to virtual) presence in the market. It can supplement its physical marketplace with a presence in the virtual world of the mobile networks. And, even more, it can add time and position pointers from the mobile marketplace to the physical world.

... and the media perspective

But the really big winners in the new network will probably be those working to ensure the smooth running of the infrastructure (with its newly built tracks and stations): the engineers, train drivers and conductors. There will need to be good road signs so that people can find their way to the station. And of course, outside the station there will have to be an army of touts telling passers-by what a wonderful idea this new train is ... Would they like to buy a ticket?

But the big question is this: How much are people or marketing companies prepared to pay for the new infrastructure, the new possibilities? Will the mobile Internet engineers be financially compensated for all their pains, or will they suddenly stand empty-handed, the proud owners of a digital network without any passengers?

Is the mobile marketplace a mobile Internet?

Media coverage of the mobile marketplace, and the way that telecommunications companies have presented both WAP and 3G mobile telephony, have tended to focus on the idea of a mobile Internet. The question is whether the mobile marketplace will actually be a mobile Internet. Early indications are that it will not be a mobile version of fixed-line access.

The Internet as a marketplace can be described as a virtual representation of the physical world – or reality. Retailers on the Internet can make products and services immediately searchable and available to online

Table 1.1 Examples of different types of personalization

Type	Implication
Name customization	You receive a letter at home that begins "Dear Mr. Donald Duck ..."
Customer-initiated customization	You leave your personal details with a supplier, who later uses them to customize information and special offers to your stated needs
Formal customization	The supplier uses official information such as income, place of abode, car ownership, and so on, and categorizes you according to these criteria. You receive offers adapted to the group that you belong to
Preferential customization	The supplier collects information about what you buy, and uses this to guess what your preferences are based on observations of other people's habits Used by Amazon.com among others

customers. In the longer term, online retailers will be able to offer everything from communications services and e-commerce to activity holidays and high-resolution video fed directly to the television monitor.

So far, there are strong similarities between the mobile marketplace and the Internet marketplace. In so far as available bandwidth allows, the mobile marketplace will also offer a range of communication services, shopping and information – either at higher or lower bandwidth. As with the Internet, actions will be personalized and adapted to customers' individual preferences. Table 1.1 gives examples of different types of personalization.

However, the mobile marketplace has a much wider reach than anything before it. It becomes synonymous with everywhere. Furthermore, the personalization will be taken into another dimension, qualitatively a massive leap for its users and in the level of sophistication of the communication: the new technology is known as positioning or localization. Positioning with the help of the GPS satellite navigation system or plotting between base stations in the mobile network will produce exact geographical positioning. Answers will thus be available to questions such as "Where is the nearest restaurant?" or "How do I get to Main Street?" As a result, marketing companies will be able to offer their customers offers specific to place. Figure 1.1 illustrates the three dimensions of the mobile marketplace.

Beyond Mobile

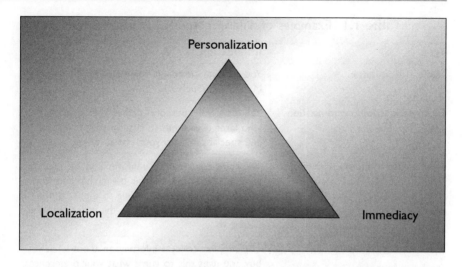

Figure 1.1 The three dimensions of the mobile marketplace

The Internet offers the possibilities of customization and immediacy – to all that are actually connected. In the mobile marketplace, immediacy is available "everywhere" and customization gains two dimensions: local relevance and position information

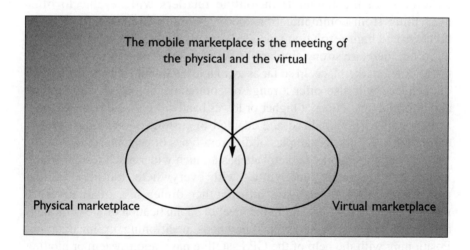

Figure 1.2 The mobile marketplace merges the virtual and physical marketplaces into something qualitatively new

What is the Mobile Marketplace All About — Really?

A marketplace is a place where buyers and sellers interact; obviously, we can talk about physical or virtual marketplaces. The perennial town square or typical boutique, are pure functions of the physical marketplace. The Internet is a pure function of the virtual marketplace. However, the mobile marketplace offers something qualitatively new from a marketing perspective. It is, in fact, a sort of interface between the physical and the virtual marketplaces (Figure 1.2).

Price comparison is well established by now as a service available on the Internet. But in the days when Internet connection was synonymous with a fixed-line cable attached to the wall, consumers could not use this service while out shopping. With the mobile Internet, price comparison becomes eminently possible. Let's suppose a consumer is standing in a shop, looking at a product that he/she wishes to buy. Simply by contacting a price comparison service and entering the relevant barcode into the mobile, it will be possible to compare the product's price with that of four or five other shops in the area. Or, if a particular product is out of stock, a similar search can be made of other shops.

SUMMARY

- The mobile marketplace is more than a mobile Internet.
- Personalization, positioning and immediacy are the cornerstones of the mobile marketplace.
- The mobile marketplace is an interface for the virtual and physical marketplaces.

Chapter 2

The Topography of Mobility, Information and the Market

Movement, information handling and commerce are three activities integral to any kind of process in human society. Every epoch is marked according to the manner in which these three key areas are interrelating. One has only to go and buy an evening paper to realize how all three are part of the most simple action: the walk to the newsagent's, a moment of information handling (*Wall Street Journal* and *New York Times*) and a bit of commerce when one digs out some change to pay for it.

The three activities may also be applied more broadly to incorporate large-scale social change. Here is a glimpse of the contemporary picture:

- Movement: Human geographical mobility has increased dramatically during the course of the 20th century. A hundred years ago, an average Swedish citizen traveled less than a kilometer per day. Today, we cover some 45 kilometers per day. Personal transportation in the 1990s increased by 2 percent within the EC.

- Information handling: The world has never produced as much information as it does today – at least if by information we mean the kind of data that can be stored in printed form, on film, or various kinds of magnetic or optical media. The Sunday Supplement of the *New York Times* contains more information than a normal individual would absorb in an entire year prior to the Industrial Revolution. A researcher at Berkeley has calculated that each year, humanity produces somewhere between one and two exabytes of new information per year,[3] the equivalent of about 250 megabytes for every man, woman and child on the planet – that is, about 250 books of text. Interestingly, it is individuals – not the mass media – that are responsible for the vast majority of this output.

- Commerce: The global economy is breaking all records. In the last 20 years, the gross global product – that is, the sum of the GDP of every country in the world – has increased by some 260 percent.[4] In the same period, the population has only grown by 35 percent. At the same time, economic systems have been restructured so that half of the world's biggest economies are actually transnational companies.

These scattered items of information are all loosely part and parcel of the development towards a mobile marketplace. Hence it may be appropriate to sketch some of the main contours of the history of movement, information handling and commerce.

Table 2.1 Work, education and entertainment – change over four epochs of human history

Type of society	Work mobility	Education mobility	Entertainment mobility
Hunter-gatherer society	Constant movement to new hunting grounds to hunt large animals over large areas	Nonexistent	Nonexistent
Agricultural society	Work in own fields, temporary travel to trade	Exclusively young aristocrats on tours abroad. Wandering apprentices. Pilgrims	Extremely limited
Industrial society	Extremely limited for most. Tied to machines in the factories	Travel to university cities for growing numbers of people towards the end of the 1900s	Travel to and from summerhouses, visits to relatives, and later charters and other tourist travel
Knowledge-industrial society	Knowledge workers carry their work with them. Work where and when you want is a reality for many	Larger groups of people study abroad. Educational travel is a possibility for the masses	Roaming by train and plane. Backpacking for the experience of it. Travel becomes a pleasure in itself

Movement

People have always been on the move. Nothing is so intimately linked with the idea of freedom, as the ability to move ourselves whenever we want to.

But if we look at human habits over a longer time period, it is apparent that human movement has changed significantly. Only in very recent times (say, the last few decades) have people begun to recreate the mobility that characterized the age of hunter-gatherers.

If we subdivide movement into movement in working life (work mobility), education mobility and leisure mobility, it becomes even more plain to see what major changes have set in over the last few hundred years (see Table 2.1).

From nomadic hunters, to agricultural settlers to global knowledge-nomads

Mankind is a wandering species. Throughout history, adventurousness and necessity are both there: nightly campfires, daily expeditions into the world.

Geographical displacement – whether a one-hour walk, a one-year journey or a lifetime of wandering – has always been an important way for individuals to express choices in life. People have always been able to vote with their feet. Freedom and mobility are often viewed as synonyms. Few people would ever give up their right to mobility – hence the prison house is a punishment.

Under the hunter-gatherer period, we shadowed the animals that we hunted and the plants that we ate. In this way we dispersed beyond our original home in Africa, and eventually populated the entire world. The first real farmers first appeared about 10 000 years ago. In various places in the world and independently of each other, they began to cultivate certain plants and domesticate animals such as horses, oxen, water buffaloes and others.

Agricultural communities became more stationary, but there was still some degree of movement. Most excursions were fairly limited, perhaps a few kilometers at most. Many individuals spent their entire lives in the same parishes.

The era of industrialization reduced mobility even more. There was a certain amount of movement to get to the factory, certainly, but once in their places of work, people had to adapt themselves to the rhythm and demands of the machines.

Only now, in the knowledge-based society, is mobility on the way up again. More and more workers have jobs that could, at least in theory, be done almost in any place and at any time of the day. Flexi-time has become more common, but home-working has not become as established as it was thought it would be, in the 1980s and 90s. "In those days we thought IT would drive development on. That's not the way it's worked out." Human needs have defined things, claims Lennart Sturesson, research fellow in Technology and Social Change at Linköping University, where he recently completed a doctoral thesis on distance working.[5]

One of the most fundamental human needs is friendship and/or fellowship. People need to meet, and to talk on a personal level. Historical data show that the level of mobility tends to correspond to the level of exchange of messages, letters, and so on. Higher levels of mobility seem to bring about higher levels of indirect communication, and vice versa. Only at times of war do other forms of communication momentarily replace travel.[6] It seems that people use information technology to a very great extent in order to arrange meetings.

One of the most important milestones in the history of personal travel is the establishment of the railway in the latter half of the 1800s. Once the railway was up and running, it was possible to link even the most isolated communities in different parts of the country. A journey of 500 kilometers was no longer a disincentive. The railway would kill off horse-driven transport, it was gloomily predicted. Instead, it rejuvenated it! Short-distance travel to and from the railway stations led to a boom in horse-drawn cabs.[7] Even by the turn of the 19th century the urban environment was dominated by the horse and carriage, becoming a source of constant irritation in the process. Hackney cabs and horse-drawn trams crowded the streets, manure piled up in the gutters, and the stables attracted swarms of flies. However, once electric trams arrived in the cities in the early part of the 1900s, the horse gained a serious competitor even for the shorter trips.

The victory parade of the motorcar did not begin in earnest until after the Second World War. In the 1950s, EC countries had no more than about 10 cars per 1000 people. Today, this figure has jumped to 450. The car became a means to individual freedom, as illustrated in Figure 2.1. Nowadays, the car accounts for six out of 10 journeys along Swedish highways. Measured in kilometers per person, its dominance is even more pronounced.

Our travel habits have altered surprisingly little over the past century, in the sense that the time spent traveling every day has been almost constant. Year after year, we seem to spend 50 minutes traveling to work. But in those 50 minutes we travel 50 times further today than at the beginning of the 1900s. The car is without a doubt the most significant factor in this development.

Figure 2.1 The significance of car ownership in the 1950s

In the 1950s, the motorcar was the most important symbol of freedom in Europe. This image of the nuclear family setting off for a picnic in the countryside symbolizes both the security and the freedom of car ownership

Source: Volvo, Historical Archive

The car has also become a safe haven for our personal lives. There, we can sing along to our favorite songs, or spend a moment philosophizing – unless the mobile phone rings, of course. The function of the car as a mental autonomous zone – a place where the driver is free from awkward or pressurizing demands – is perhaps the biggest obstacle to phasing out commuting by car.

On the other hand, it is not only the car that offers the possibility of mental space. Swedish Railways' (SJ) advertising campaign shows people daydreaming while the landscape passes by outside. The most exciting journeys are within ... is the slogan of the campaign. Travel time is dreamtime, in other words.

The time we spend traveling is often also devoted to work. In the first six months of this year, sales of laptop computers rose by more than 50%.

The Topography of Mobility, Information and the Market

Almost 20% of Swedes own or plan to own a laptop computer; among academics, the figure is closer to 40%.[8] The USA, Western Europe and Asia/Pacific follow a similar pattern with increases of over 30%. Portables seem to be gathering steam as users give weight to the advantage of mobility over resources for applications.[9]

It is common to see the matt glow of a laptop computer on the night-time train. And aircraft are full of people aiming to get through those reports they should have read long ago.

Traveling also gives rise to a fair amount of dead time. Traveling is waiting. This is an opportunity for advertising, film and other entertainment to gain a foothold in our attention. As a result, complicated alliances have been built up between companies that move people from place to place – and hence are in a position to control their attention – and companies within sectors such as advertising, entertainment, media and education.

Another function of journeys is to create fellowship. It is often said that whoever travels, always has a good story to tell. Literature is full of travelers' tales, from *Gulliver's Travels* to *The Hitchhiker's Guide to the Galaxy*. The journey of enlightenment and education is a long-established tradition, and journeys to exotic locations provide a precious entry ticket to the world of the cosmopolitans.

Other kinds of friendships are created around sailing boats, classic cars and motorcycles. In this context, the journey from A to B is largely irrelevant. The road itself is the purpose. Preferably it should be full of dangerous bends, at least if one consults the large numbers of men in their forties, most of them drunk on freedom, who dominate the world of motorcycling in Sweden. And even in the severest of bends they will gladly let go of the clutch handle to give a wave to soul brothers passing by on the other side of the road.

Information handling

For the greater part of human history, the majority of people have lived in a very local environment with information in short supply. Most of them did not meet more than a hundred or so people in their entire lives, and the vast majority never even learnt to read.

In the last few centuries, and not least in the most recent post-war period, the flow of information has changed radically. This has also affected our capacity for handling information and possibly even our way of thinking. Possibly we have become wiser.

In the transition to a mobile marketplace, with constant access to all the knowledge in the world, the ability to select, sort and evaluate information will become a decisive factor.

From paucity of information to constant excess

> On an average weekday the *New York Times* contains more information than any contemporary of Shakespeare's would have acquired in a lifetime.[10]

The information environment has changed radically through history. The majority of the Western population is now no further than a few clicks away from the Internet – the world's greatest ever "omnium gatherum" of knowledge, entertainment and garbage. Currently we are attempting to handle this with a cognitive organ that for most of its developmental history has existed in a world without written language. Modern man, *Homo sapiens*, evolved in Africa some 200 000 years ago. The earliest remains of written language are about 5000 years old. It is not surprising that the kind of information overload to which we are subjected on a daily basis, occasionally makes one's brain hurt.

Obviously the brain is affected in the short term by the kind of information it is exposed to. If it were not, we would not possess a memory at all. But what is the relationship between our modern, complex lives and the evolution of the convolutions and functions of the brain?

One man who devoted a great deal of thought to this was Alfred Russel Wallace, who formulated important precepts of Darwinian theory before even Darwin had had time to do so. The two men corresponded and had a great deal in common. However, on one important matter their ideas diverged strongly: Darwin believed that the human capacity for thought had evolved by means of natural selection. Wallace, on the other hand, maintained that it was a gift from God. The proof of this, Wallace stated, was that the minds of savages were as perfectly made as those of civilized people. "Natural selection could only have endowed savage man with a brain a few degrees superior to that of an ape, whereas he actually possesses one very little inferior to that of a philosopher."[11]

Wallace's meaning, in other words, was that if Darwin's reasoning on natural selection had applied to neurology, hunter-gatherers, living more primitive lives as they did, would have more primitive brains than the rest of us. This not being the case, the brain thus had to be a divine gift.

Wallace's big mistake was obviously his total misjudgment of the hunter-gatherers' lives. Theirs is a society founded no less than any other society on social cooperation, which places extremely high demands on the brain. The complexity of social relationships is well illustrated by the case of computers – while they excel at chess, they have never made good friends and they certainly are not good at social interaction. It is the intensity of information that humans are so good at handling. Given the weak physique of humans compared with animals, obviously our species had to be clever if we were to survive at all.

First of all, then, one can draw the conclusion that Westerners are not the first people in history who have to process overwhelmingly complex information on a daily basis. The nature of the information flow has changed, however. Early people could live out their entire lives on the savannah without meeting more than a hundred people. Modern people might meet a hundred people in a single morning. Conversation outside the church on Sunday mornings was an important newsgathering activity right up until the advent of newspapers towards the end of the 1800s when the press became a mass media. In the 1900s, radio, television and the Internet have multiplied the information flow and created a whole new news environment. These changes produce clear neurological changes that are measurable in the longer term.

The so-called intelligence quota, or IQ, is a widely accepted measure of human intelligence. The extent to which it actually measures human intelligence is another matter, on which there is wide disagreement. Yet whatever the merits of the system, it is interesting that the average human IQ has been rising ever since the tests were introduced at the beginning of the century.

Countries with military service commonly put their young male recruits through IQ tests. In the Netherlands, all 18-year-old recruits are tested using an IQ test known as Raven Progressive Matrices. The young men's fathers and grandfathers were also given the same test. In the last half-century, the results have been improving by about three points per decade. Similar data has been noted in about 20 other countries. This means that a person of average intelligence today would have a higher IQ than 84 percent of the population in the 1950s. Nutritious food, better education and smaller families might also have contributed to the development; however, many psychologists believe that changes in the information flow, to which we are constantly exposed, have played a decisive role. "My best guess is that the increase is caused by changes in the visual environment. TV, computer games, toys and advertising contain elements that increasingly resemble IQ tests," says Ulric Neisser, Professor of Psychology at Cornell University.[12]

Every new generation adapts to the prevailing information flow in a way that can be measured using IQ tests. Other repercussions are more difficult to describe statistically. Marianne Frankenhaeuser, Professor Emeritus of Psychology at Stockholm University, has for many years been conducting a study into the effects of information stress.[13] It is a malady that impacts people in different ways. Frankenhaeuser comments: "Those best placed to filter out unwanted information are the well-educated. They often have a foundation of knowledge that gives them an ability to recognize contexts and meaning." However, no one has an infinite capacity for handling information. Once the flow grows beyond our capacities, we begin to make poor decisions; particularly if the information is contradictory. There are interesting patterns here: "We begin to indulge in wishful thinking. We simply decide that the information means what we want it to mean," says Marianne Frankenhaeuser.

Another effect of the increased information flow and, at the same time, the increasing inaccuracy of the information, is that people are choosing to make decisions on qualitative information rather than pure facts. People are not capable of interpreting more than three or four facts in order to make a rational decision. If there are more than seven or eight facts to take into account the decision will not be fully rational. And if there are even more factors to consider, the quality of the decision-making is significantly reduced. At such times, friends and advisors become more important in the decision-making process; or people fall back on their emotions – perhaps their feelings for a brand – to make a purchasing decision.

Meanwhile, people have a growing awareness of the dangers of information overload. Bodil Jönsson became a focus for this idea with her book *Unwinding the Clock*,[14] which, like so many detective stories, was widely read and eventually translated into English. One of her central theses is that the brain, much like any machine, needs running in. People need to be acclimatized before they can engross themselves in an issue. Hence the scattered flow of information must occasionally be switched off, thus giving the individual time to think. Thinking takes time. And it takes time to separate salient pieces of information from the junk. This is something that both senders and receivers have to appreciate in the emerging information society.

The fact that thinking takes time relates to the actual function of the brain. For a new thought to be established, new synapses have to be built up, connections have to be made with already existing structures. These processes take a great deal of time, and at times of intensive thought in areas where previously no great attention has been lavished, the brain consumes more energy than the rest of the body put together. Martin Ingvar, Professor of Neurology at Karolinska Institute, Stockholm,

The Topography of Mobility, Information and the Market

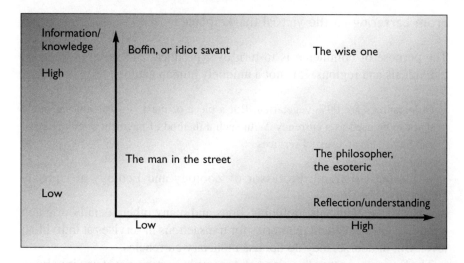

Figure 2.2 Wisdom presupposes both high-level access to information and a high degree of reflection

believes that we have to devote at least six or seven minutes to a new issue if we are to remember it at all.

Information on its own, then, does not necessarily lead to development and "intelligence." Information must be coupled with reflection. Data and information that we reflect upon and analyze can become knowledge, and knowledge tied to experience can eventually turn to wisdom, as shown in Figure 2.2. The information perspective on history shows that humanity as a whole has gone from information impoverishment to overflow. The real question, or at least the significant question, is whether we have got any better at using the information available to us. In a mobile information society, what are the implications for reflection and deep knowledge when news and information are presented in two-second headlines?

Commerce

Commerce has existed in all ages. Of course, it has been subject to change – from local bartering of animal skins and salt, to the symbolic commerce of the virtual marketplace. The mobile marketplace is a new and evolutionary step, enabling people and machines, regardless of their geographical position, to exchange services.

From bartering to the virtual marketplace

The purpose of commerce is to turn over products and services between individuals and regions. It is not a uniquely human activity.

> Chimpanzees are 99% vegetarian. But a piece of meat is a valuable resource that can be used as a currency. With such a method of payment, a chimpanzee can buy status and sexual services

says Staffan Ulfstrand, Professor of Zoology and Ecology at Uppsala University.[15]

Apes – like elephants, dolphins, lions and many other socially oriented animals – have excellent memories for transactions in services at individual level. But of course, they do not use monetary currencies.

The human economy has also got its roots in bartering systems involving physical contact between the buyer and seller. Nuggets of precious metal were first alluded to as currency in the Old Testament, but the first real coins were minted in Asia Minor – about 600 BC – in the Greek cities of Ionia and the kingdom of Lydia.

The monetary economy gave rise to new market patterns. Commerce flourished via intermediaries – merchants. From this, the step was not so enormous to what we still have today – city streets full of shops. Of course, many revolutionary events have occurred in the intervening period.

In Northern Europe, for instance, the invention of a seaworthy and hardwearing merchant ship (known in Dutch as the *cogge*) was a prerequisite for international trade, emerging by the middle of the 1100s. Under the protection of the Hanseatic League, the *cogges* plied the seas for half a millennium. The Hanseatic League was a political and economic federation of German merchants with cities in the North Sea and Baltic region, which effectively secured trading monopolies.

During the 1200s and 1300s the number of central European cities grew at an unprecedented rate; and, indeed, unprecedented ever since. This led to expansive levels of commerce; with the economic activities of cities growing ever more specialized.

Other important occurrences in the development of commerce were the discoveries of the New World and trade contacts with China and India. Commerce was largely conducted using monopolies and privileges.

The demise of the privilege system did not come about until the industrial revolution, which started in Great Britain at the end of the 1700s. The mass market began to emerge, and it was not long until Henry Ford made his legendary promise that consumers could buy a car in any color they liked, as long as it was black.

Continued economic development has hailed a period of disintegrating borders. In Europe, the fall of the Soviet Union and the establishment of the interior market have stimulated this. Deregulation in a host of areas including currencies, has also contributed to the current situation. In later years, a growing volume of e-commerce trade has tested national regulatory systems to the full. The problem was perfectly expressed when last year a French court ordered the US Internet portal Yahoo! to make some of its web pages inaccessible to French citizens.

The reason for this was that Yahoo! was holding online auctions of medals, flags and Nazi memorabilia from the time of the Third Reich. According to French law it is illegal to trade in racist symbols. Similar cases in nations all over the world are now causing cracks in commercial legal frameworks.

Repeated changes in the basic assumptions of trade throughout history, have had major implications for the relationship between buyers and sellers. Before the time of mass markets and bulk buying, commerce was something transacted in the physical marketplace between two people who knew each other well. In a novel written by the Swedish writer Göran Tunström, the character of Stellan Jonsson is a latter-day version of an anachronistic merchant. He always keeps a jar of figs for Pastor Cederholm, a tin of asparagus heads for Mrs. Bergström at the post office, beef consommé for the dentist, Apelkvist, and so on for the whole community. This is a case of one-to-one, personalized marketing. Even in larger companies, in some cases, fairly sophisticated information was held on the customers. For instance, the sewing machine pioneer Isaac Merritt Singer is alleged to have collected information about families that bought his machines, so that the company could make another approach once the daughters were old enough to get married.

The emergence of the mass market at the beginning of the 1900s changed the entire focus. Suddenly there was a surfeit of buyers. Customer information became less relevant; as did the concept of adapting services to the customers' individual needs. The real attention was directed at production. Ford's Model T is a clear example of this. During its 19 years in full production, it accounted for half of the world's production of motorcars, and it was only available in black (Figure 2.3). In those days, systems for amassing information on the customer were simply of no great interest to the large manufacturers – unless some moral strictures were involved. In Sweden, the state alcohol monopoly, Systembolaget,

Figure 2.3 The Model T Ford

Long before the age of mass customization Henry Ford promised
that everyone could buy cars in any color they liked, as long as it was black

Source: Pressens Bild

kept careful notes on the drinking habits of its customers. Not, however, in order to increase sales.

With increasing affluence and growing competition in the market, the customer soon became a scarce resource. The seller thus needed to get to know the customers and tailor products and marketing towards designated groups, possibly even towards particular individuals. Store cards and credit cards linked to databases made it possible to monitor purchases, build up customer profiles and detect new consumption patterns, which gave renewed emphasis to the question of personal integrity in the information society.

Virtual commerce (or e-commerce) made it both easier and more difficult to take care of customer relations. It became all too easy for the customer to leave electronic footprints while also taking on the dogsbody

Table 2.2 The development of the markets from barter to an excess market

	Characteristics	Means of exchange	Meeting
Barter	Exchange of goods and services in the local market	Goods, silver	Physical meetings between buyers and sellers
Trading in goods	Emergence of standardized trading goods sold over large geographical areas Merchants become middlemen Strict regulation of commerce	Money	Physical meetings between middlemen and buyers
Mass market goods	Mass-produced goods sold to large groups Excess of available customers Large industrial companies dominant Production capacity is a key factor Free commercial activity, regulated international trade	Money	Physical meetings between middlemen and buyers
Excess markets	Mass customized goods and services sold to increasingly narrow groups Deficiency of customers Brands grow in importance Customer relations become crucial Free commercial activity, international trade deregulated	Money, e-money	Physical and virtual meetings between physical and virtual sellers/middlemen and buyers

job of scanning product availability in inventories before making his/her purchase. And more difficult, in the sense that the marketing of local services and opportunities ran absolutely counter to the geographically unspecific structure of the Internet.

In the US advertising market, the largest in the world, companies spend almost US$100 billion every year on local advertising.[16] To get their hands on a slice of this cake, many companies now offer services that can determine the locality of an individual surfer. It thus becomes possible to reach anonymous Internet users with advertising relevant to his or her local market. The most rootless of all markets, the Internet, is becoming local.

With the emergence of the mobile marketplace, and with built-in intelligence and communication functions in goods and services, society is moving towards a market *where not only people but also products become buyers and sellers*. Among the technical visions are intelligent refrigerators that order new food supplies independently of their owners, cars that order extra horsepower for their engines, or washing machines that order servicing – all of this occurring in an online environment.

The mobile marketplace

From a long-term perspective, we can see that the mobile marketplace in many respects is a natural, evolutionary continuation of a long historical process. As we have already said, people are mobile, social beings. For hundreds of thousands of years we have been a communal species with a need to move unhindered across massive areas while at the time retaining the social structure of the group. In the last millennium, people with the opportunity to do so, have traveled further and further for purposes such as commerce, the acquisition of knowledge and adventure. Over the same period the availability and flow of information and access to knowledge has increased by fits and starts. The invention of writing and printing as well as the school system and mass media have led to a broader, deeper world consciousness and an improved ability to handle large amounts of complex information.

The appearance of the mobile marketplace will enable people to "keep the flock together" wherever we are in the world. We will also gain access to all the knowledge of the world and all of the world's markets irrespective of where we are.

Presented thus, the mobile marketplace is a natural extrapolation of a historical curve established long ago; and quite simply, may be viewed as something that makes life a little easier. The big question is assuredly not

The Topography of Mobility, Information and the Market

whether in a generation or two we will regard the mobile marketplace as a necessity. It is rather this: Will today's adult generation, which has grown up without mobile telephones or the Internet, see it as something natural or even desirable? Will those of us who are older than 30 perceive permanent Internet connection as a stressful necessity, a phenomenon that shatters the peace and stops one from thinking a single thought from beginning to end?

> **SUMMARY**
>
> - People have always moved around; mobility is natural.
> - The knowledge-based society creates an increased need for movement.
> - Man is a social creature; the mobile phone helps the flock retain social cohesion.
> - From information scarcity to information overflow.
> - Wisdom presupposes access to information, but also time for reflection.
> - From physical to virtual trading places, in which machines buy from machines.

CHAPTER 3

The Individual as a Mobile Creature

> No, we are not nomads. We do not have the right qualities to be successful nomads. Basically, we are social individuals who gain our information and ability to survive out of meaningful social relationships. (Professor Martin Ingvar)[17]

The individual is a social creature, then, but also a mobile one. Our mobility does not only relate to our physical bodies. We are also mentally mobile and are adept at migrating between various mental states, various identities.

As we move around outside the home and workplace we need to make use of a whole range of services. The mobile telephone quickly becomes a success all over the world. But what kinds of mobile services are most relevant, and which of them are unique to the mobile marketplace? And how does state of mind, our mentality, affect the sorts of services we ask for?

Mobile archetypes, states of mind

In a rather cut-and-dried way it might be said that mobile services are useful to two kinds of people: those who are short of time (and believe that the mobile phone can save them a bit of time) and those who have plenty of time on their hands, and wish to make "dull-time into fun-time" to quote Soki Choi, the Swedish-Korean mobile commerce entrepreneur.[18]

Such a cut-and-dried breakdown of human nature instantly translates into a huge success for the mobile marketplace, as long as prices can be held at acceptable levels. More and more people are short of time, and young people in need of fun and diversion are also on the up.

However, the perceived advantages of the mobile marketplace may not be quite so simple. If we look at mobility in a broader perspective, we find that it can be subdivided into different categories. It is one thing to talk about the mobile marketplace while away, another while at home. There is a big difference between finding oneself in an unknown place and walking down the street where one's mother lives. In an unknown location local information is useful – shops, maps, and so on. When we are on home ground, this kind of information is not what we are looking for. Besides, there are at least two kinds of long-distance travelers: the amateur who only travels sporadically and the professional traveler. For the latter, travel is not associated with any real pleasure, hence the importance of utility services. The most important thing is to be able to do one's job in whatever location one finds oneself. For the amateur, the journey is usually made with some sort of recreational activity in mind, and the need for mobile services is probably a good deal lower.

When traveling or moving around, we are either static or in motion. When we are in the street, sitting in a plane or in a car, we are in motion. On the other hand, we are static (although the definition is somewhat vague – after all, we are actually static while sitting in the plane) when working as consultants with a customer or while staying in a hotel in an unknown city. Our need for mobile services and our ability to use them are highly dependent on whether we are static or in motion (Table 3.1).

Another important dimension, which determines how prepared we are to be disturbed by incoming signals such as telephone calls, e-mail, advertising, is our mental state or mood. When waiting to board a plane at the airport, or even while just walking down the road, we are fairly likely to be in the mood for "getting things done." To an ever-greater extent we

Table 3.1 Dimensions of mobility

Goal	Close: local mobility	Distant: long-range mobility
Purpose	Entertainment: leisure mobility (the tourist, the idler)	Utility: working mobility (the business traveler, the shopper)
Frequency	Temporary: temporary (the charter tourist)	Frequent: professional mobility (the golf player, the project worker)
Speed	Stationary: on site in another location. Waiting	Movement: going somewhere else
Mental state	Negative: waiting, focusing on setting off or arriving	Positive: relaxation, inner focus. Activity

have a tendency to do more than one thing at a time, while waiting for something or in transit: such as send e-mails, make a few phone calls, plan the agenda of some imminent meeting or make arrangements for dinner. Of course, we might also be focused on nothing – or as Erich Fromm would call it, "being." In the first mindset we would probably opt to use all kinds of utility services such as web surfing, using e-mail, referring to an almanac, making notes or short letters. In the second mindset we would not be interested in any kind of mobile services. Perhaps we would prefer not to be disturbed at all, but rather turn off the mobile and create a personal safe haven.

Both of these alternatives are relatively stable mental positions. For as long as we are unwilling to be disturbed by outside signals or thoughts to force us to deal with reality, we are mentally able to stay in a contemplative state. To retain a work mode mindset is a relatively simple matter. However, there is also a mental halfway house, which, unlike the other two stable mental positions, relates to an unpleasant suspended emotion – namely the state of waiting for something to happen.

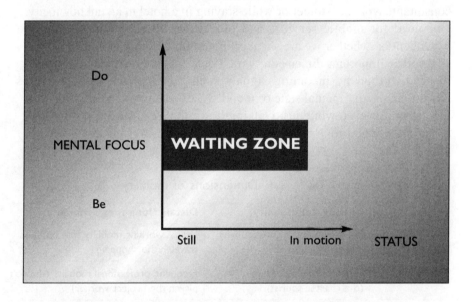

Figure 3.1 The dimensions of mobility

We can be mobile both while stationary and in movement. Most important of all in terms of our receptiveness and interest in messages is our mental focus. While waiting, we are most open to impulses from the external world that take us out of the negative waiting state

When we are waiting we are focusing mentally on the actual waiting. At such moments, time seems to creep by at an infernally slow pace. At worst, we start getting irritated at the bus that never seems to come, the plane that is delayed, or the train that never arrives. Our sole interest is in arriving, or getting away. The waiting zone – for instance, a queue – is therefore a highly interesting moment for those who wish to attract our attention (Figure 3.1). Yet even the digital world consists of waiting zones caused by downloading or other digital delays. The frustration caused (techno stress in IT jargon) would probably have amazed earlier generations. They would have been amazed at the intense frustration of spending hour after motionless hour staring at a glass screen.

To sum up, there is evidence to suggest that in moments of waiting when we find ourselves suspended in inaction, we are highly receptive to outside impulses that take our attention away from the actual waiting. We are looking for promises, as Professor Martin Ingvar put it, promises to take us away from an unpleasant place.[19]

Mental mobility and games of identity

> The essence of life in a post-identity society isn't that people have no identities at all. Rather, it's that people have more identities than they know what to do with, that identities change or cease to mean what they once did. (Walter Truett Anderson)[20]

People are undergoing psychological changes that are connected to the ever more kinetic society that we live in. It is easy to see how each generation grows better at adapting to increasingly movable lives, but it is not always clear in which direction this development is heading.

The postmodern society offers a dizzying array of identities and roles, and any vaguely adventurous individual gets accustomed to dipping in and out of roles and structures in a way that has not been possible at other times in history. The digital networks reinforce the possibilities of role changes and experimentation in identity. Once we cut loose our communication and symbolic language, do we experience this as liberation or something that produces angst? In other words, how does this mental mobility affect the postmodern individual's view of him or herself?

Regis McKenna is fond of describing media as an extension of the human psyche; the media is a sort of vehicle of everything we know, everything we think and dream, and even the transactions we make.[21] We are becoming more flexible as consumers. We switch rapidly between various consumer perspectives, consciously and unconsciously, as we navigate through the media buzz. Our consumer behavior is affected to a very high degree by our attitude to time, to the media, to change and, indeed, to how we handle information. Modern man might be described as Homo Consumentus, with three subgeneses that we switch between at will: Homo Consumentus Politicus, Homo Consumentus Ludens and Homo Consumentus Effectivus.

Homo Consumentus Politicus

Many individuals among younger or middle-aged generations are practically born in the supermarket trolley – in a world burgeoning with products. At the same time, there is an increasing awareness of the poverty that prevails in many parts of the world. "Think of the children in Biafra" was a commonplace parental exhortation in the 1970s to recalcitrant children who left food uneaten on their plates. Consumerism colors practically every aspect of contemporary life, whether it's a case of browsing in New Age stores or examining the best available pension funds.

Consumers are aware of their new power, and are prepared to vote with their feet and impose boycotts, to stake out the road and express their views through consumption.

Homo Consumentus Ludens

The playing human grows happier as a result of shopping. This is now also verified by research. Clearly, play is a serious matter. People seem to be playing in a rational manner – emotions and play are at the foundation for the interpretations we make, and behind the curiosity that helps us develop. People have higher expectations of the sort of content with which they wish to fill their lives. We require a certain number of "kicks" per second. Advertising has become socialized. We play with lifestyles and expect to be addressed in an appropriate way – a way that fits the role we are currently playing – at the appropriate time.

Homo Consumentus Effectivicus

The effective individual strives to simplify life and gain time to do whatever he or she chooses. Technology is regarded as an empowering factor, and the individual is always asking for new ways of making daily life simpler. Consumers are becoming more and more sophisticated, with enormous demands on quality. This sophistication also encompasses a strong feeling for appropriate formats. From an information perspective, there are also compelling arguments to suggest that individuals will be looking to technology to help them reduce the need for decision-making, and sort and filter the information flow.

Different kinds of mobile service

As we have seen, people have been mobile creatures from their very inception. It was only once people became agriculturalists, and later, during the era of industrialization, that settled life became the norm – mobility in work or living was severely curtailed thereafter.

What tasks can we carry out while mobile?

Earlier, we discussed the unique dimensions of the mobile marketplace – that is, the combination of positioning and immediacy. Thinking in terms of time-and-place-specific services is one of the ways of categorizing the sorts of services that are available.

So what kinds of actions might we want to take while on the move? Obviously we want to talk to other people: to be contactable and to be able to make contact with others. The rapid development of mobile telephony is a clear example of this. The culture of small talk has definitely impacted on the world of the mobile phone. Small talk and contact are time-sensitive. They happen in the present, not later on. Increasing numbers of people, not least young people, want to indulge in small talk in the form of text messages. Currently, they do this in the form of SMS messages.

Another kind of text-based small talk that has emerged in the 1990s is e-mail. E-mail is also used for business communications, with content playing a central role. Having access to e-mail, also in the form of longer messages and attachments, is a distinct advantage for the professional user. Appropriately enough, most people would like to use improved mobile phone networks to send e-mails.

Being able to track down like-minded people in a new place, or when on the town, is another interesting service, not least for young people or people with clear-cut interests. "What people with similar interests to my own are in this town?" or "...can be found within a 500 meter radius?" At its most sophisticated, this type of service can be developed into a kind of dating service. When you pass someone in the street that conforms to your "dating profile" – and vice versa – the mobile gives a beep, and the two of you have the opportunity of meeting in the real world. A first-generation version of this kind of matchmaking service, known as "Lovegetty," already exists in Japan.

There are many other kinds of basic positioning services that could be of consumer interest, such as "How do I get to Main Street?" or "Where is the nearest McDonald's?" These types of service are often specific to time and place, as shown in Figure 3.2.

Occasionally, pure information services can also be specific to time. While shopping, for instance, we occasionally get impulses from the environment, requiring an instant response. "What is the name of the band that did this song – can I order it?" Or, "What other shops around here sell these Salomon skis?" The option of adding still or moving images to a phone call might also add certain advantages.

Other time-critical information to which one wants access everywhere might include: internal company information, access to intranets (without

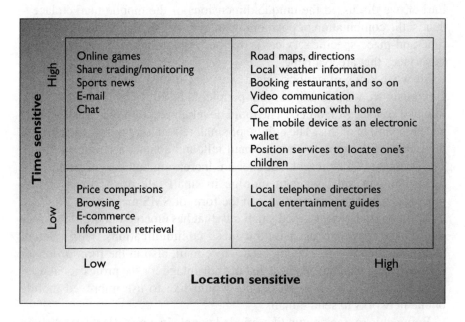

Figure 3.2 The mobile marketplace: time- and position-critical services

The mobile marketplace will be filled with services that are time- and position-critical to a lesser or greater extent. Services that are both time- and position-critical are using the potential of the mobile marketplace to the full

Table 3.2 Various mobile services and willingness to pay

Type	Example	Willingness to pay
Communication	E-mail, SMS, chat	Low
Transaction services	Ticket booking, banking	Relatively high
Entertainment	Music downloads, network games, film, radio	Variable
Information services	News, cinema listings, weather, directory enquiries, encyclopedias, price comparison	Variable
Position services	Road directions, restaurant tips, locating one's children, finding friends in the vicinity	Variable
Security services	Turning electricity on and off, checking to see who is ringing the doorbell	Variable

having to carry heavy computers), stock exchange information, racing information (for instance, making a bet on a race that has already started), news summaries, and so on.

Finally, there is a range of services in areas such as security: checking who is ringing the doorbell, or activating or deactivating the burglar alarm at home from a remote location, or other instances of this kind.

Another kind of service that has very little to do with mobile phones, but which has become popular in Finland as a pure mobile service, is the possibility of making small payments – using the mobile phone as an electronic wallet – such as the mobile phone bill, feeding parking meters, and so on.

In other words, there is a wide range of possible services. Broadly, they can be categorized in six different groups, as shown in Table 3.2.

Empty or full channels

From the section above we may conclude that consumers will make use of the mobile networks in one of two ways, and we can view these as two types of channels or marketplaces – empty or full.

Empty channels or markets are meeting places that users fill with their own content. These meeting places are like a film set without a director or actors. An empty meeting place is like a café where the guests create the noise in the form of discussions around the tables. On the other hand, a full meeting place (or channel) is like a theatrical performance or lecture.

Many of the services mentioned above are empty channel services. SMS, e-mail, telephony and chat are all examples of this. Interactive online games are a borderline case, either empty or full depending on how they are delivered. Many of the most successful Internet services have been of the empty kind. Chat, e-mail and forums are examples of services that have radically affected communication patterns and information handling. Downloadable music has also proved a successful format; its sheer accessibility has probably been an important factor in its success, but also the possibility it offers of consumers creating their own personalized compilations of songs.

Of course, full channels also have their success stories. But the question remains: Where will the principal demand be? In full or empty channels?

A new interpretation of McLuhan's prophesies

The whole discussion about empty or full channels has its source in the ideas of one of our time's most enigmatic media theoreticians. When we come to consider the growing impact of the media on modern man, established concepts such as "cold" and "hot" become ever more relevant as interactivity continues making inroads into traditional media. Marshall McLuhan attached great significance to the levels of engagement of various media – and held that this would be crucial to their development in the future.[22] The comparative hot and cold terminology was taken from the world of jazz. Hot media are characterized by a tendency to dissolve; they are vigorous and self-contained and leave little for the consumer to become involved with. Cold media do not dissolve, and leave more for the consumer to fill with his/her own imagination. Through this, a higher level of engagement is created. Currently, there is an excess of hot media in society. Hence, interest is growing for cold media. McLuhan made a further distinction, which he termed "light on" and "light through" media.

Light on media were characterized by a light source directed at a dense surface that did not allow any of it to pass through. These types of media have always existed. However, he continued, people had always been fascinated by media that were light through – in other words, lit up from within like a monitor. The electronic media had better prospects than any other media of becoming multifaceted and multidimensional. Many of his coinages from the 1960s were perceived as difficult, pretentious and contradictory. He viewed himself as an explorer rather than a teacher. During early computerization in the 1970s he began to muse on the changes he was seeing, and what the effect of these would be on consumption and social patterns. Only now in the early 21st century can we begin to properly understand the meaning of his observations. We find ourselves in the midst of a media development that gives renewed importance to his thoughts and makes him highly relevant.

One of his most famous utterances (and perhaps also the most misunderstood) is: "The media is the message." By this, he meant to say that the actual use of more and more media was more significant in terms of social change and human behavior than the actual information content of those media.

Historically, whenever a new medium has entered the fray, we have always sought to refer to and relate it to the known world. The important thing here, according to McLuhan, was that the choice of metaphors often indicated a limited imaginative scope in the pioneers of a new medium. The feature film needed fluid storytelling, had to be within one and two hours in length and was structurally based on the novel. In the early days of radio, the new medium was viewed as a wireless telegraph and was hence rather flatly dubbed "the wireless" in the English language. There was no clear conception of how to communicate with the unseen, broad masses. Perhaps only the Sunday service achieved a sort of mass transfer – from one to many. The program presenters sounded like preachers or telegram readers. With the arrival of television on the scene, the structures of radio were initially superimposed onto the picture on the screen. For a long time, television was reminiscent of radio with a visual image. The *raison d'être* of video was our desire to use the television as a content source whenever we wanted to – and no longer be bound by whatever the television channels were showing. Perhaps only now the media development has become so immensely complex that we no longer know who the Internet content providers of the future will be, we have started talking about the wireless again, this time in the context of mobile phones – once again showing a certain lack of imagination in our insights about developments in the coming years. It seems likely that we will once again fail to predict the real effects of this new medium in society.

One of McLuhan's central theses was that one generation invented and formed the tools, and that those tools formed the next generation. The relevance of this to contemporary life is clear. For instance, the inventors of the Internet seem not to have fully understood its implications; whereas the generation that has grown up with surfing, computer games and mobile instant communications have been formed and conditioned by it.

"Tomorrow's people don't actually read the newspapers, they get into them every morning like a hot bath" is another well-known comment of McLuhan, referring here to the revolutionary changes in media structures that he believed would occur within a not-too-distant future. Many people could not quite see the significance of the statement, but in the light of digitization, an explosion of channels and information overflow, his meaning now seems eminently clear. Perhaps around the next corner is a total immersion in information – more or less like a hot bath. The question is, who will control this information flow? Are people capable of browsing and sorting this immense flow of information? Will we allow advertising into our private bathtubs, in return for services such as packaging and sorting of the information? Will we even notice the burgeoning amounts of advertising slugging it out for our attention?

The mobile purchasing cycle

What are the implications of mobile commerce? Is commerce really mobile at all?

To understand the mobile marketplace from the buyer's perspective we first have to look at the mobile purchasing cycle. Whether for a product or a service, every purchasing decision begins with an impulse. The mobile marketplace is characterized by the ability, as a result of mobility, to act on impulse. Evidence seems to indicate that in an age of increasing information and choice we are going to be governed by impulses – in fact, we will consciously allow emotions to take over the decision-making process. An advertising billboard, a magazine advert or a special offer via the mobile might trigger the impulse. Equally it may be triggered from within – a sudden desire to do something that has been forgotten.

The impulse to action is followed precisely by action. If there is not the opportunity of taking the action, or even the awareness of it, chances are high that the impulse will simply remain as an impulse. Nothing will be done. The mobile phone helps us realize there is something to be done – as well as the practical means of actually doing it. Hearing a song that we like, we can immediately find out what the song is and perhaps even order it. If we see a job advertised in the classifieds section, we can immediately access information about the company. Or in the event of a car advert that catches our fancy, we can e-mail the owner and ask to be contacted.

The next step in the purchasing process is about closure – or, actually making the order. Most goods that are offered on the Internet are actually bought in a high street environment. The purchases are made on the basis of information available on the Internet. The importance of e-commerce for general retailing is thus higher than sales figures seem to indicate. The prognosis for the mobile marketplace is still unknown. For certain kinds of goods, the market share will be small, and in others considerable. Who knows, perhaps on the way home from work people will check the contents of the fridge and arrange to have a delivery waiting on the doorstep when they get home.

As for digital services such as downloadable music and games, these have a more definite and immediate closure.

If costs are low, the threshold goes down and the likelihood of impulse buys goes up. Very few people buy 20-meter motor cruisers on impulse, whereas when it comes to ice cream they do.

When we refer in this book to mobile commerce and the mobile marketplace we mean the whole chain from impulse to information search to closure to delivery (Figure 3.3).

The uniqueness of the mobile marketplace is that it creates a short cut in the path from impulse to information to purchase. The entire journey is just a matter of a few clicks, irrespective of where we are. All in all, this is a world in which impulse has more importance attached to it, and this adds to the power of emotion.

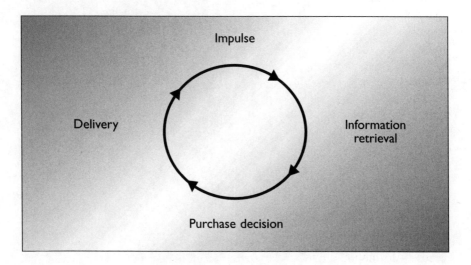

Figure 3.3 The buying cycle

In the mobile marketplace the process from impulse to delivery
(in digital form) or completion (for physical products)
can be reduced to pressing a button

SUMMARY

- Where, when and why we are mobile is decisive for the demand for mobile services.
- The mental state of mind is crucial both for the demand for services and the receptiveness to advertising messages.
- Waiting is associated with something unpleasant.
- Mental mobility leads to games of identity.
- Empty and full channels have different functions.
- The mobile marketplace is a short cut through the purchasing cycle.

PART II

Preconditions, Driving Forces and Trends

We have now seen that the mobile marketplace is emerging in a context where incremental and slow social change, combined with the odd giant leap, are reshaping society and individuals. In many senses the mobile marketplace seems to be a natural progression. We have also seen that mobility generally and mobile services in particular are vague concepts.

We are now going to probe deeper into the possibilities created in the transition to the knowledge-industrial society, in terms of new technology, new institutional guidelines and new values.

We start by looking at the interaction between different forces of social change, and thereafter in orderly fashion run through major issues in the areas of technology, the individual, institutions and the economy.

Part 1

Preconditions, Driving Forces and Trends

Chapter 4
Technology, Institutions, People and the Economy

Anyone purporting to know all the reasons why the mobile marketplace is suddenly just around the corner, is being economical with the truth. The technical ability of turning an idea into reality is one reason; the willingness of the telecommunications industry to take the plunge is another. Nor do we know with any degree of precision what the real implications of this new marketplace will be. Perhaps it will be little more than a mobile version – although a little more sophisticated – of the congregation staying on for a chat after Sunday Mass, or a sort of Speaker's Corner. Maybe it will develop into an electronic butler – a "Guide on the Side" – to help us with minor services or a bit of navigation in the physical or nonphysical world. Perhaps it will occasionally keep us entertained when we are bored, and provide a bit of educational back up when we least of all expect it. All these things will collectively improve our communication with the world around us.

Fundamentally one can talk about four cornerstones, each one of them necessary if the mobile marketplace is really to take off. Some of them have already been constructed, others are unfinished and not yet firmly in place.

The first of these is the *technical prerequisites*. Only now do we have the technical ability to make the mobile marketplace a reality. Up until today, we have only been able to fantasize about mobile commerce, mobile electronic guides, constant Internet connection, freedom of mobility in the labor market, and so on. Using third-generation mobile telephony, improved PDAs and screens, and so on, some of the fantasies postulated by the visionaries can actually become a reality.

The next cornerstone is *institutional prerequisites*. There have to be rules of play, organizations and players that provide support or at least do not obstruct the developing mobile marketplace. There have to be "carriers" of new technology and the right preconditions have to be in place in order to establish new technology. One of the big obstacles for the breakthrough of the digital economy, for instance, is the many still unresolved issues relating to intellectual property rights (IPR).

The third cornerstone is the *human prerequisites*. There certainly seem to be many convincing indications that the mobile marketplace would serve a whole range of human needs. But many questions remain – for instance, how are people in practice going to use the possibilities of the mobile meeting place?

Important social changes often arise as a result of the interplay between technology, people and institutions. These three cornerstones enable the fourth cornerstone to come into play – that is the new economy, the new value multiplier – that is, *new economic prerequisites* (Figure 4.1). Technical

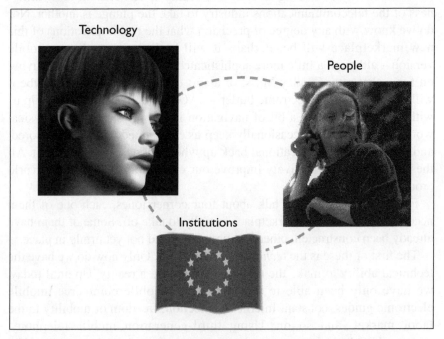

Figure 4.1 The interplay between technology, people and institutions
The relationship between technological (here illustrated by Ananova.com), human and institutional prerequisites creates the foundation for new economic realities
Source: Ananova Ltd 2001. Reproduced by permission. All rights reserved

possibilities and human wishes often do not correspond with what is financially viable or even achievable at all. If there are no ways of making real money out of the mobile marketplace it is likely to end up as a half-empty digital churchyard.

Social change as a game of push and pull

The interaction between new technology and human needs, values and visions, institutional preconditions and economic realities, lead to a type of social development that might be characterized as a reactive game of push and pull.

The Dutch historian Fred Polak wrote on Western society from a very unusual point of view, publishing a macro-historical work on European developmental history, that attracted a great deal of attention, in the early 1950s. The title of the English translation was *The Image of the Future*. Polak's starting point was to examine the importance of visions and utopias in history, and their role in helping propel societies forward by qualitative leaps. He concluded that the ability to give form to what had not yet been experienced, the ability of the individual to dream and have visions of how life should be, were crucially important for continued development. But he also stated that this development could be described as a relationship between the forward push of the visions and the resisting pull of history.

If we apply these ideas of Polak to the scenario of the future mobile marketplace, we would do well to initially ask where the inertia lies, and where the forward movement. What drives development down a certain path, as a consequence of historical pressures? What functions as a storm anchor, reducing the ship's speed on the sea of history? Or, on the other hand, what dreams and visions are sufficiently compelling to focus development towards the mobile marketplace?

To begin with we should question whether there really are any real forces, or motivations, that lead towards the creation of a mobile marketplace. Is the mobile marketplace the answer or response to any deep-seated human needs or dreams? Can the vision of the connected individual move humanity a step onwards? Or is the mobile marketplace just a sordid attempt by the global telecommunications industry to profit from its capital?

What is immediately apparent is that large groups of people seem to nurse a dream of a life in freedom, a life unfettered, a life beyond the limitations of physical space. For some, this dream takes such strong expression that, in an almost Gnostic spirit, they dream of the disintegration of the body and a life lived as a virtual representation on the net. But for those of us who entertain less advanced spiritual notions, for the more conservative "thirty-somethings" among us, it is enough to have access to information and communication wherever we are. This dream of freedom is intensified by the descriptions of technical utopians of the promised land – the possibilities inherent in mobile technology. Technical innovators and entrepreneurs also hold a great deal of power by constantly seeking out better and more innovative technical solutions, and always advancing the frontiers of technology. The question is if these dreams are strong enough to wield any real power. Can they really work as a center of gravity, a center of attraction for the complex reality of the mobile marketplace?

Historical forces are about the effect of decisions made in the past on the evolution of the future. Whatever we do today, or did yesterday, will have repercussions on the future. Sometimes these can be more decisive than we suspect. In chaos theory, the idea is sometimes taken to its logical extreme – chaos theorists sometimes state that the beginning is *everything*. The important thing therefore is to choose the right way. The Internet is a good example of this. Because from the very beginning information on the Internet was provided free of all charges, a culture was quickly established in which it was impossible to levy a charge on information. Similar decisions in the mobile marketplace will have repercussions that are every bit as crucial. If, for instance, the mobile phone operators choose to keep prices high for GPRS so that consumers are not deterred later by the very high cost of UMTS, this might lead to very slow growth in mobile data traffic. Of course, it is important to add that while this *might* happen, it might also *not* happen.

The regressive and backward forces holding back development and slowing down the progress on the historical journey, are often present in the form of constitutional barriers (a lack of copyright protection, for example), technical obstacles (too few standardized solutions) as well as people's mental limitations, which stop them from seeing the true potential of any given situation.

Among the forward-moving, positive forces in the mobile marketplace, we must certainly include the desire of the telecommunications industry to start earning revenues from investments already made. It is primarily mobile phone operators that are referred to here, but also systems suppliers. With the impending transition to third-generation mobile telephony, the incentives are considerable. The gigantic investments being made effectively mean that the mobile phone operators will do everything to increase traffic in the 3G networks.

As we have said, there are also obstacles and problems that impede the pace and focus of development. Some of the limitations are purely technical, some tied to issues such as standardization and consensus, and then of course there are financial constraints. Finally, it is the users that might present an obstacle, at least from the point of view of the industry as a whole. Do they actually want the new technology? And how much are they prepared to pay for it?

Technology and people – a necessary whole

Technological innovation and the implementation of new technology are closely related to human behavior. Below, we are going to take a closer look at this relationship, how human expectations and preconditions affect the speed with which new technology is introduced.

Technical breakthroughs produced by gradual development

New technology can be new in different ways. Occasionally it is based on pure innovation and representative of something entirely new, but it is more common in an evolutionary form. Occasionally, new uses are found for existing technology, and occasionally different technologies are combined in new ways. There are five distinct forms of technical innovation – from the small steps of transformation to the revolutionary leap.

Transformation is the most common form of development. Transformation means that a technology by means of a series of small steps successively turns into something else. New subsystems are added on without altering the original functions. Examples of transformation are cars, which are equipped with ever more sophisticated technology, or cameras, which are becoming increasingly electronic.

Expansion means that existing technology is expanded by means of add-on systems, so that the original functions are broadened. The machine is capable of more. Expansion might also be described as an integration of new functions in an existing technology. Examples of expansion are teletext, TV cards in computers, video conferencing via PCs or mobile telephones.

Fusion is when new products are based on two or more earlier technologies, and when the new products are more than the sum of their parts. Examples might be portable computers equipped with GPS, digital nautical charts, or the Nokia Communicator (a computer and mobile telephone). Possibly, one might also classify the multimedia computer as another example of fusion, when equipped with TV, video, voice mailbox, DVD player and data communication. The Internet is a fusion of computers and networks. The extent to which a particular development is representative of expansion or fusion often depends on one's perspective.

Substitution is a well-known process. A classic example of substitution is the motorcar, which became a substitute for horse-drawn carriages. Other examples include the transition from propeller planes to jet turbine-powered aircraft, mechanical clocks giving way to electronic time-keeping, or the move from NMT to GSM in mobile phones.

Innovations are news pure and simple. Innovations do not complement existing technology, but tend instead to create entirely new markets. The video, Walkman, facsimile, mobile telephone and synthesizer are by this definition all examples of innovations.

The question, in other words, is whether we should regard the mobile Internet as a pure innovation without any obvious competition from existing technology, or if, when all is said and done, it is merely a fusion (between the Internet and mobile telephony) or substitution (from a fixed-line to a mobile Internet). In the case of the third-mentioned, we would expect mobile Internet use, to some degree, to replace fixed-line Internet use. If it is more a case of pure innovation, the new technology would eventually mean that the Internet were available in all sorts of new places where previously it had not been available. If we consider the example of the Walkman: after its launch, music was mobile, music was everywhere.

For new technical systems to achieve a breakthrough, however, it is not enough for them to merely work as they are supposed to. Much more than that would be needed to set a whole new mobile marketplace in motion. Table 4.1 illustrates the various layers or platforms that have to be in place before a new technical system can achieve a breakthrough.

Once the technical platforms are in place, there is a further requirement for standardized handling of data and file formats, as well as security, encryption, and so on (the second layer). Add to this the necessity of fully functional infrastructure – that is, reconstructed networks – as well as users (the third layer). Gaining acceptance among users is not something to be taken for granted. People – that is, potential users – have to have a chance of trying out the new technology and getting used to it.

The growth of home computer ownership in Sweden, for instance, was facilitated by the fact that the government made computers tax-deductible. A century ago, to give a push to the sales figures of "automatic carriages" (that is, cars), the car manufacturers were offering driving courses! To really bump up the sales figures of mobiles, Nokia, Ericsson and the mobile operators should probably start running WAP and GPRS data-users' courses.

Finally, real advantages must be perceived in order to get users to accept new technology. New technology has to represent something *significantly*

Table 4.1 For new technology to break through, four subsystems have to be in place: technology that works, systems for data and information handling, infrastructure and functioning content

User Service	Information	Buying service	Buying experience	Buying product
Service	Train times, sightseeing and so on	Advisory services	Games, entertainment and so on	Music/video, tickets
Acceptance and *Building Infrastructure*	Acceptance To use, to trust	Dissemination Number of terminals and so on	Knowledge How advanced are the services that can be delivered?	Infrastructure Coverage, quality and so on
Data Information	Data format HTML, XML, and so on	File format MP3, Quicktime and so on	Presentation format WAP, browser and so on	Positioning Encryption, payments and so on
Technology	Terminal Mobile telephone, PDA, and so on	In/out Voice recognition retina display, pen on screen, and so on	Technology for data communication GPRS, UMTS and so on	Positioning GPS, and so on

new and different to make people adopt it and use it in their lives. This was, and continues to be, the main problem of the digital television. Because it does not represent anything significantly new, there is no incentive for the consumer to go digital. New technology has to represent something new, and be truly useful.

The S-curve and driving forces in various phases

In other words, the emergence and market penetration of new technology is a complicated process. From a marketing perspective, the form this complexity takes is that the initial establishment among users is often very slow. In the pioneering phase small groups of enthusiasts with considerable technical skills and little of the usual consumer interest in things such such as content and user-friendliness, drive the process. This group experiments and, because of its high technical demands, helps the developer iron out any remaining technical wrinkles in the system. This initial phase in some instances can continue for many years – for as long as there are no standardized formats for data/information, and infrastructure and dissemination are limited, there will be no appreciable growth in user numbers. To use a celebrated example, the Internet was used at this level in academic and research establishments for more than 10 years, without anyone outside the small user group knowing anything at all about the phenomenon.

Eventually, the new technical systems enter a colonizing phase. With this, broader groups of users come into contact with the new technology, users who see great potential in it, but who place much more stringent demands on aspects such as functionality and user-friendliness. If we stay with the Internet example, the transition to the colonizing phase began with the development of HTML and the emergence of what was initially called the World Wide Web (www).

The colonizers lay down the foundations for mass appeal. However, the mass market will only come into play once prices have gone down substantially or a watershed number of people are using the system, thus almost making it a necessity for everyone else to have access to it.

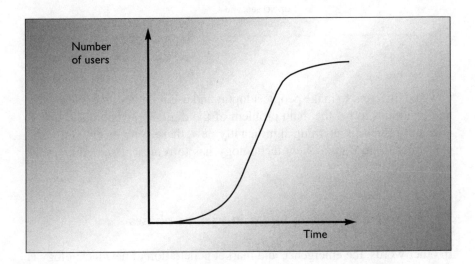

Figure 4.2 The introduction of new technology in a market often takes the form of an S-curve

Eventually it becomes more difficult to reach out to new users. Enthusiastic technophiles and visionaries have already jumped on the bandwagon, as have the majority of the more conservative users. The only people left are the stragglers, considerably more difficult to catch in marketing terms. And with this, the process finally starts slowing, as illustrated in Figure 4.2. User numbers are no longer increasing as rapidly as they were during the mass-market expansion phase. Hence, the industry's focus turns to further technological development and fine-tuning of existing systems and technology.

If we look at the development of mobile telephony in Scandinavia, the 1980s belonged to the enthusiasts. At the end of the 1980s and early in the 1990s, user groups were broadened first with the introduction of the NMT900 system, and thereafter GSM. Ever larger groups of professional users were acquiring mobile phones. For business use it became axiomatic, but private ownership was still comparatively low. This secondary development, the mass-market phase, did not take off in earnest until the latter half of the 1990s, with hefty discounts on mobile phones and prepay packages. Now, at the beginning of the 21st century, when the majority of

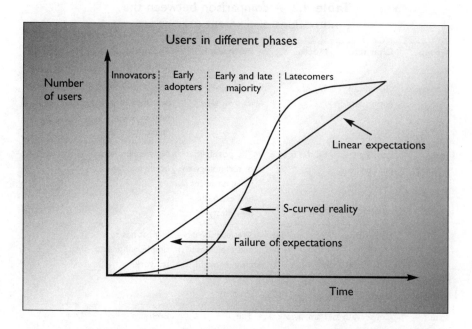

Figure 4.3 Interplay between expectations and actual development

The interplay between expectations and actual development often lead to a phase where reality does not conform to expectations – and with this, rejection of the possibilities of new technology

young people and adults have got a mobile phone, the expansion has started to abate. Many of our senior citizens will probably never buy a mobile phone, just as they will never purchase a computer or sign up for an Internet service provider (ISP). Hence total saturation in terms of mobile phone ownership is expected within a few decades.

A central question in the mobile marketplace is if the S-curve might be compressed or short-circuited by developing new services with broad mass-market appeal. A comparison between WAP and the Japanese I-mode suggests that this might be possible. The superior user-friendliness of I-mode in combination with its broad range of easily accessible services, many of them interesting to the youth market, have thrust I-mode into a mass-market phase almost immediately while WAP is still struggling in the gloomy depths of the pioneer phase. Not even the hard-bitten enthusiasts have the patience and energy to look for interesting sites.

In other words, as a market expands, new groups of people suddenly become customers. Many companies that have achieved success in the

Table 4.2 A comparison between the motivations of different user groups

Group	Character	Motto	Motivation	Technical demands	Willingness to pay
Innovators	(Technical) enthusiasts	Try it!	Technology for its own sake	Low. Are often prepared to help develop applications	Low
Early adopters	Looking for possibilities, visionaries	Be the first!	The possibilities of technology to make advances and establish new lifestyles	New applications – useful and fun	High
Early majority	Pragmatists	Wait and see!	Are functionality and usefulness clear enough? Follow the majority	High. It has to be simple, and it has to work!	Relatively high
Late majority	Conservatives	I am skeptical	The pronouncements of authority	Even higher!	Low
Latecomers	Skeptics	Never!	Resistance to the new	Can it ever be good enough?	Nonexistent

early stages fail to reach the mass-market because of new demands placed on the products by customers. Ericsson's success in the infancy of the mobile phone revolution, and its subsequent disappointing results in the face of greater consumer sensitivity to price and design, is one well-known example of this. Companies that really want to "hang in there" for the entire duration of the cycle, have to understand the mechanics that apply in the various developmental phases. One of the most important lessons is the nature and motivations of the different user groups. This is the key to short-circuiting the S-curve – that is, to quickly enter the mass-market phase. Table 4.2 illustrates the nature and motivations of various user groups.

The tyranny of expectation

The biologist and systems theorist Gregory Bateson defined information as "a difference that makes a difference." By this he meant that we could appreciate something as information if it could be placed in a context that was useful to us. On the basis of this, we could interpret, gain understanding and draw conclusions.[23]

As beings, people are information seekers, but also seekers for meaning. Our intellectual drive is to place things in their context. We also have, almost physiologically, a defined need to draw conclusions about the likely implications in the future of changes (or differences) we observe in the present. Unlike other living beings, as the neurologist David Ingvar demonstrated so clearly,[24] humans create scenarios.

Through our endeavors to intellectually capture changes and then extrapolate them into the future, we also have a tendency to enlarge small phenomena. Small, slow changes in the S-curve are quickly scaled up. In other words, we have a tendency to exaggerate the speed and potency of any given change, as illustrated in Figure 4.3. Once it becomes apparent that the actual development does not quite reflect our expectations, there is a risk of a corrective backlash. The result is an all-too-big credibility gap between "what could have been" and "what is."

Take, for instance, Internet share valuations. In autumn 1999 the Internet phenomenon exploded onto the Nasdaq register and other technology-based stock markets. The expectations were enormous, particularly for e-commerce. But already by March 2000 it became clear that very few e-commerce companies would be able to perform according to expectations. After the high-publicity collapse of boo.com, the air went out of the market like a punctured bubble.

However, it is when the market experiences these sorts of conditions that the truly serious mistakes are made. Because the reality has not lived up to hopes and expectations, people make the mistake that the whole thing is finished. The call goes out: "Sell your shares!" Instead of coming to terms with the fact that the process has not quite reached maturity, and then settling down to wait for another lift-off in a few years time, people are likely to find another "golden calf," on which to base another set of expectations and extrapolations. The about-turn of the venture capital community from B2C to B2B to m-commerce within the space of a few years, is a typical example of this kind of flightiness. At the time of writing, the focus seems to be shifting towards technology companies.

In order to be able to surf the breaking wave of growth, one needs to understand the trends and have the knack of preempting the right moment.

SUMMARY

- More than technical platforms are needed to establish the mobile marketplace. Widespread acceptance and practical services are just as important.

- Introduction of new technology takes the form of an S-curve, but our expectations for the breakthrough of the new technology are linear and often higher than the actual experience.

- Early adopters of new technology have different demands and expectations than the broad mass.

- To accelerate the introduction of new technology, the S-curve has to be short-circuited.

Chapter 5
Technology

Let us define technological prerequisites as the sum of existing technology combined with people's access to that technology. In this way, people are seen as active constituents in the whole technology equation.

But it needs to be clarified right from the start that the events of technical revolutions cannot be foretold with any degree of accuracy. The uses to which new technology is put can only be vaguely perceived, or guessed at. And often – all too often – our guesses are wide of the mark.

For instance, when Richard Trevithick built the first steam-driven locomotive in 1803, he knew that it would revolutionize the mining industry. He was, after all, a mining engineer. He did not realize, however, that within 50 years trains would be moving people here, there and everywhere, in fact all over Victorian England. Nor did he perceive that trains would play a crucial role in the expansion of many towns into cities, as well as making the bathing holiday a national institution.

Similarly, Vint Cerf and other Internet pioneers did not realize at the beginning of the 1970s that e-mail would be the true killer application of the net.

In this section we look at the various technical prerequisites of the mobile marketplace. First, the technical platforms for achieving a mobile marketplace – that is, 3G networks, local wireless networks and positional marketing. Second, how new technologies are being established and what the expectations are for growth and development. There is a short summary of the technology that lies at the heart of the mobile marketplace. Finally, we make some forays into the remarkable world of software agents.

Moore's law – and Metcalfe's

There are two laws – commonly known as Moore's and Metcalfe's – that are deeply significant in the development and proliferation of new technology.

Twice as good every 18 months

Gordon Moore, one of the founders of the chip manufacturer Intel, discovered in the early 1960s that microchips were becoming twice as powerful every 18 months. Ever since, his observation has been something of a landmark in the technology industry, and Intel has used the 18-month principle as a founding business concept. If one looks at the IT sector in its broadest sense, it quickly becomes apparent that development in many areas (not least data transfer speeds) has been even quicker. If we assume a development pace of doubled performance and halved prices every 18 months, and if we then apply this model to the motorcar industry, a Volvo S60 in six years' time would weigh less than 100 kilos, use 0.005 liters of fuel/km, have a load capacity of several tonnes, a top speed comparable with that of a modern fighter plane, and a cost of less than US$2000.

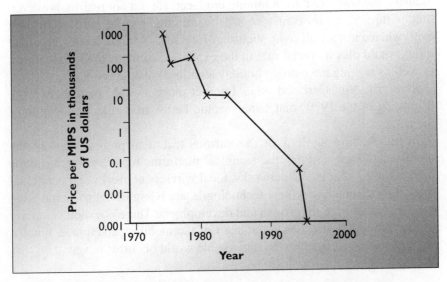

Figure 5.1 The cost of computation power calculated in MIPS (millions of instructions per second)

Source: New York Times

Network benefits are exponential

Metcalfe's law is about the benefits of networks. The greater the number of people using a network, the more useful the network. And this utility grows rapidly, for the reason that the number of new "pair relationships"[25] increases exponentially when new users join up. In a group with n participants, the number of pair relationships is $n \times (n-1)/2$. If the number of users increases, for instance, from 5 to 10, the number of pair relationships increases from 10 to 45 – that is, more than four times. If the number of users increases from 100 to 1000, the number of pair relationships increases from about 5000 to 500 000 – that is, a hundredfold increase.

This founding principle is one of the reasons why communication technology often has initial problems in establishing itself. Once user numbers exceed a certain threshold, development can quickly become explosive. This was the case with both fax and e-mail.

Metcalfe's law also explains why it is so difficult to maintain large groups. Once numbers become too large, subgroups form and tend to pull in different directions. Groups of 12 (that is, 66 pair relationships) seem to be the upper limit, a magic threshold beyond which it is not advisable to go. Someone once said: "No group should have more than twelve members. Jesus opted for twelve, and that proved one too many."

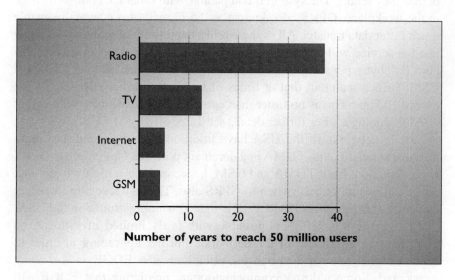

Figure 5.2 New technology breaks through at a quickening pace
Source: New Scientist, 21 October 2000

New technology established more quickly

One of the repercussions of general technical innovation in combination with growing affluence is a speeding up of technical development. Radio, introduced in the 1930s, took 40 years to reach 50 million users. The GSM telephone took four years to achieve the same result, as illustrated in Figure 5.2.

Mobile telephone networks

What are we talking about when we use terms such as the mobile Internet and third-generation mobile telephony? The technical foundation for our current mobile systems was laid down in the early 1990s, with the introduction of GSM as a standard in many parts of the world. GSM networks are package switched, which means in practice that we pay for the time we are connected rather than the amount of data transferred.

GSM networks are currently being upgraded to work more efficiently as vehicles for data transfer. In Europe, these upgraded networks came into service in 2001, and by autumn the first mobile phones were flooding the market. The new networks based on GPRS technology are package switched, which means that users are constantly connected, but only pay for the amount of data they send.[26] The system is comparable with Ethernet, commonly used in the workplace. GPRS, also known as 2.5 G, will also eventually enable much faster data transfer. All of the world's experts are quibbling about how fast the service will actually be. The theoretical number being used is 115 kbps (kilobytes per second) – that is, about 10 times quicker than traditional GSM. Critics maintain that at times of high user traffic, speeds will not exceed 10 kbps. This is no faster than current speeds achievable with earlier GSM technology.[27] For further details see Table 5.1.

Mobile networks in the USA have traditionally subdivided into more than one standard, with TDMA in a dominant position. Upgrading to GPRS is possible both from TDMA and GSM, hence making it highly likely that American operators will switch to GPRS also. This is backed up by recent evidence, showing that GSM/GPRS is indeed winning ground in the USA.

The transition to GPRS is possible with fairly limited investment, no more than an upgrading of the networks. Further upgrading in order to achieve higher rates of data transfer, often known as EDGE, is also being considered. Some industry commentators are predicting that EDGE will become the dominant technology in mobile data transfer, as opposed to UMTS or other 3G networks.

Table 5.1 Highest theoretical data transfer speeds of various mobile data communications. In practice, speeds will be considerably lower

Technology	Theoretical data transfer speed kbps	Introduction	Type of net
GSM	9.6	1992	Circuit switched
HSCSD	57.6–115		Circuit switched
GPRS	115	2000	Package switched
EDGE	384–473	2002	Package switched
3G	384 (while mobile out of doors)	2001 (Japan)	Package switched
	2000 (indoors)	2002 (Europe)	
4G	54 000 (indoors)	2010?	Package switched Seamless integration of various types of networks

A significant and heavy argument in favor of UMTS, quite apart from improved speeds, is that UMTS uses frequency space more efficiently than GSM and its evolutionary forms (GPRS and EDGE). Efficiency is a crucial issue here, as the overall frequency space set aside for civil use is limited. EDGE also uses higher frequencies, making the range of the senders more limited than GSM systems. GSM, in turn, has a smaller range than the NMT450 network used in the Nordic countries in the 1980s. In rural areas, short-range signals are obviously disadvantageous, because network expansion is expensive. However, in urban areas short-range signals become an advantage – areas can be divided into cells, and more calls simultaneously handled. In short, capacity is improved.

While mobile networks are developing, other, to some extent competing, solutions are also appearing. The digital radio standard DAB is already up and running commercially. Unlike the mobile networks, based on one-to-one communication, the DAB network like any other radio network is a communication medium intended for many receivers. This makes the system suitable for real-time information aimed at a mass market, such as updating of company information, news broadcasts, share prices, and so on. By adding a back channel, for instance through a GSM link-up, the same interactive possibilities are added as in currently available shopping channels in digital television. In other words, alongside the digital terrestrial networks (DAB and digital television networks) there is also the possibility of creating a mobile Internet.

Wireless local networks

Wireless local networks are already becoming a reality for many companies. The laptop user here and there has already invested several hundred dollars or so, to enable a wireless connection with a printer or the Internet, while working with a laptop computer on a sofa or hammock. And more and more people are buying cordless headphones or speakers for a hi-fi unit. The critics of UMTS expansion are championing wireless local networks as a credible alternative to 3G networks. The former manager of Nokia's GPRS systems, Hannu Kari, has expressed doubts that the speed of UMTS will exceed 144 kbps with all the inherent bottlenecks in the system, and believes that GPRS will not achieve much more than 10 kbps. Rather, he views the Internet as the hub of a network where local wireless networks are tied to GSM-, GPRS- and UMTS-nets.

Wireless local networks are already reaching speeds of between 5 and 10 Mbps (megabytes per second), which is many times better than what UMTS will achieve even by the most optimistic predictions. The technology is cheap. Its only requirement is a telephone and computers able to switch between local networks and GPRS networks.

The development of wireless local networks has picked up speed very quickly during 2001. The market is segmented in three parts: wireless networks in the workplace, in public services and in private homes. Wireless local networks (wireless LANs) are already being set up in residential areas and places of work, in hotels, railway stations and airports. There are plans to turn whole cities into wireless hot spots. British analyst BWCS predicts there will be 120 000 public W-LANs globally by 2006.[28] By 2002 mobile operators are planning to integrate W-LAN in their mobile subscriptions.

The real boost for wireless networks in the home is expected in 2003, when new industrial standards come into force.[29] These will feature a transfer speed of 54 Mbps – 25 times quicker than present-day broadband – making it possible to view television or video content on any portable device in the bathroom, in the garden, in the attic or, for that matter, in the hammock ...

Total integration between different types of wireless networks, so that receivers automatically select the network with the best transfer rate at any given moment, is part of what is currently being dubbed 4G (Figure 5.3). In a technical sense, the development of systems and devices is already well under way, but no real breakthrough is expected until about 2010.

Technology

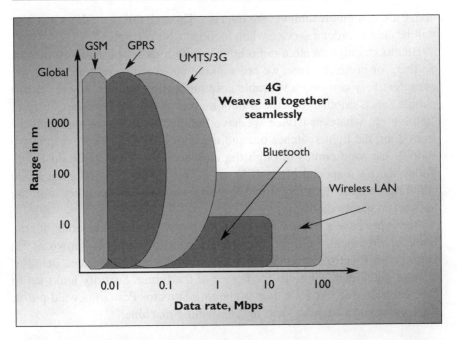

Figure 5.3 Transfer capacity and range of various wireless networks

Bluetooth – wireless micro-networks

There are many different kinds of local wireless networks. One of the standards developed by Ericsson is known as Bluetooth. Inspiration for the name came from Harald Bluetooth, a Danish Viking chieftain whose main achievement was to unite warring tribes of Vikings. It remains to be seen whether Ericsson's Bluetooth will manage a similar feat. So far, a few thousand companies have signed up to join the Bluetooth clan.

In spite of delays, Bluetooth is now up and running and more and more phones and gadgets are equipped with the technology. The prognosis from various research institutes is also largely positive. According to Merrill Lynch,[30] by 2005 there will be a couple of billion Bluetooth chips in circulation – in other words, one in three people will have some kind of Bluetooth-equipped gadget.

The idea of Bluetooth is to wirelessly link up various gadgets – anything from a mobile phone with an earpiece, to a control panel in the home regulating the washing machine, lighting, central heating or hi-fi equipment. Machines, in other words, will be equipped with Bluetooth equipment to facilitate wireless communication with other apparatus in the home and, above all, communication via the Internet. Through this, a washing machine will be able to order a service when feeling "a bit run down."

Bluetooth chips create a radio bubble with a radius of about 10 meters. So that, for instance, when we are walking about with a mobile phone, we are effectively carrying a movable communication bubble. Whenever this bubble meets another Bluetooth bubble, our mobile has the capability of "talking" to whatever device we have bumped into. As we approach the front door, the light switches on, and the door greets us. As we step into a shop, a message comes through the earpiece: "Welcome, Donald Duck. Today there's a special offer, 10% off all our stock ..." And even while standing outside looking into the shop window, we can try out the networked games on display.

Big hopes have also been placed on the penetration of Bluetooth into the home environment. In 1999 Ericsson set up a joint project known as E2 Home with Electrolux, the white goods manufacturer – the idea being to develop intelligent solutions for homes of the future. Security looks set to be a central issue. As E2's then Managing Director Per Grünewald put it, "No one wants a Melissa virus in the washing machine."[31]

Positioning

As we have already said, positioning is a crucial part of the equation in the overall development of a mobile marketplace. Positioning makes it possible to pick up geographically-related information, thus enabling companies to send out geographically-related advertising. In terms of the technology required for localization or positioning, this will be given a big boost by developments in the US market. The Federal government (US Federal Communications Commission) has stipulated that by October 2001, 67 percent of all emergency calls must be capable of being tracked within a margin of error of 125 meters. Once positioning begins to be a broadly available feature some time in 2002, the issue of privacy will certainly come under discussion. There will be large numbers of people who take issue with the fact that certain services will give friends, the boss and spouses the ability to always know where they are. Governments in Europe and the USA are looking into the question, but no decision is expected within the EU until 2003.[32]

Technically speaking, positioning can be achieved in many different ways. The particular technical system that takes precedence over the coming years is not yet clear. The most likely scenario is that several different technical systems will work in parallel.

Technology

Technology today and in the future

In a technical sense, the mobile marketplace may be described as a fusion between the Internet and mobile telephony. Hence, one of the factors determining how rapidly the mobile marketplace is established will be the availability of mobile phones and the Internet. The USA and Northern Europe are in the front ranks of Internet usage, with around half of the adult population having access to the Internet. In countries such as South Korea, Singapore, Hong Kong, Australia and New Zealand, user levels are similarly high. Northern Europe is also in a leading position when it comes to mobile telephony. Around 75 percent of adults own mobile phones.

However, development is blindingly fast. In countries such as the UK, Italy and Germany, where mobile phone use was markedly low just a few years ago, figures are now well on the way towards or beyond Scandinavian levels. During the latter half of the 1990s, reality frequently exceeded all prognoses when it came to mobile telephony use.

Figure 5.4 shows three groups of countries: the USA, the Scandinavian countries and Japan/Southern Europe. In countries with relatively low Internet usage such as Italy and Spain, the mobile Internet may for many people come to be viewed as an alternative to fixed-line Internet connection.

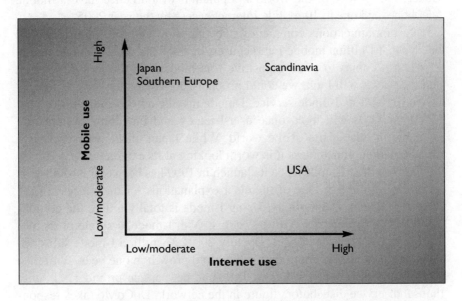

Figure 5.4 Mobile phone and Internet use in different groups of countries

This has largely been the experience in Japan, where DoCoMo (the company behind I-mode) looks set to become the biggest mobile Internet operator in the world. More than 70 percent of mobile phone users are also using mobile data services, that is, the Internet.[33] But the question of long-term development remains. Even today we can see how interest in mobile Internet use has led to increased fixed-line connections in Japan. In Scandinavia and the USA, the situation could not be more different. With most users accustomed to broadband or the convenience of large-screen Internet surfing, the mobile will probably be perceived as a complementary and rather poor cousin of fixed-line access. However, at the same time users will demand mobile services – their fixed-line Internet habits will migrate into the mobile environment. In conclusion, it may be seen that fixed-line and mobile Internets provide mutual support – demand for one increases demand for the other.

There is a broad range of predictions from various research institutes on future user numbers for mobiles, and fixed and mobile Internet statistics. By the first quarter of 2001 the number of regular Internet users was 429 million according to Nielsen/Netratings; home access was dominant among users. In the same period, the total figure for mobile phone ownership was 727 million, according to Ovum. Prognoses are pointing towards a figure of 1–1.4 billion by 2003 and around 1.7 billion by 2005. Accenture makes the judgment that 500 million mobiles will have Internet access by 2005.[34] Hence, one in four of the world's population would hence have a mobile telephone, and one in 10 mobile Internet access by the year 2005.

Telecommunications companies like Ericsson are more optimistic and believe that 1 billion mobile Internet users by 2005 is a more accurate figure.

When it comes to finding positive role models for the mobile Internet and the way it will look, we could probably do worse than look at Japan's DoCoMo with its I-mode service. I-mode may be a good indicator of how mobile services will look, once user-friendly and functional systems are available in other parts of the world. While Japan in many respects is a unique market, nonetheless it is worth looking at its experience.

The growth of I-mode since its launch in 1999 has been extraordinary, as shown in Figure 5.5. One of the clear explanations for the success of the system has been its simplicity. Using I-mode is totally unlike the complicated WAP procedures in GSM systems. With I-mode, a single press of a button accesses a host of sites. By July 2001, 27 million I-mode subscriptions had been set up, more than a threefold growth in one year. With this, DoCoMo is now Japan's biggest ISP (Internet service provider). Twenty thousand service distributors figure in the network. DoCoMo takes responsibility for adding the cost of service purchases to the users' phone bills, charging its own mark-up fee of 9 percent.

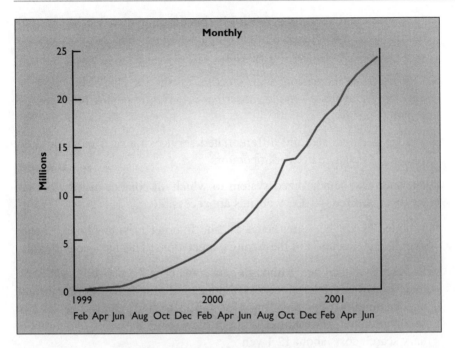

Figure 5.5 Number of I-mode users
Source: DoCoMo

The main take-up has been in the field of entertainment, with more than half of the traffic directed at such services. In general, Japanese consumers have shown a strong interest in new gadgets and entertainment services such as network games and karaoke. In the case of I-mode, particular success with games has been noted in younger age groups. However, more than half of total users are more than 30 years old, and the availability of productivity services is very high.

It is far from self-evident that the success of mobile entertainment is going to be as significant in Europe and North America as it has been in Japan. However, in preparation for the launch of the 3G networks some operators are investing strongly in entertainment and consumer services. Other analysts, like Merrill Lynch, maintain that professional users will power the development of the mobile Internet.

Among the success factors for I-mode, we might include the following:

- New data charges. A fixed price of 300 yen (US$2.5) per month for I-mode access to 500 official sites. There are also further expansion packages, also at fixed price, chosen by many consumers.

- Easy payment methods. DoCoMo handles payments with a small administration fee levied on the companies (9 percent). Seventy percent of its users subscribe to this service.
- Call traffic is increasing, with positive results for DoCoMo's revenue streams. An average I-mode subscriber is currently paying 6000 yen per month (US$50).
- I-mode personalized and differentiated services make it more unlikely for users to change network operators.
- An open and standardized system to which all content distributors are easily connected, 40 000 websites are accessible.
- I-mode has not been marketed as "the Internet in your phone," hence avoiding the creation of the wrong expectations. This has been crucial.
- Package switched net. Although data transfer is fixed at only 9.8 kbps, there is constant Internet access. Connection time is therefore minimal, and customers only pay for information actually sent. The cost of sending e-mail, for instance, is about 4.2 yen per message, and a dictionary search costs about 12.1 yen.
- Last but not least, because of NTT's monopoly on fixed-line telephony and Internet access, both of these are priced at uniquely high levels. Mobile telephony and I-mode have therefore become financially more attractive.

I-mode is now expanding into 3G licenses. In autumn 2000, DoCoMo bought into Dutch KPN Mobile and Huchison 3G UK Holdings, owner of one of the UK's 3G licenses. I-mode has also invested in AT&T Wireless in the United States, bought upwards of 40 percent of AOL Japan, and instigated a joint venture with Walt Disney's Internet Group to use Disney's well-known figures in mobile networks.

However, I-mode is not acting alone in the market. J-phone, using a similar system but with better color, had over 10 million users by July 2001. EZ, which uses WAP and achieves a baud transfer rate of 64 kbps, has 11 million users. The implication of this development is that Japan now has in excess of 50 million mobile Internet users.

Nonhuman customers

Not all potential customers are of a human kind. As we have already noted, at the same time as everything grows smarter more and more mobile customers will be electronic.[35] Dr. Keiji Tachikawa, president of NTT DoCoMo, only half-jokingly predicted that that there would be some 360 million subscribers by the year 2010, and only one-third of them people (Table 5.2). A typical user will be a soft drinks machine announcing that it needs restocking, or a lorry notifying its owner that it needs extra horsepower for the next uphill trip, or a dog letting its owner know where it is.

Fairly soon, people talking with people over the phone will be dwarfed in volume by machines communicating with other machines on behalf of their human owners. This is the view of Paul Saffo at the Institute for the Future in Paolo Alto, California:

> In future, it's not going to be your family hogging the line when you're trying to make a call. It'll be the fridge putting a call through to the washing machine.[36]

This particular example may not be entirely accurate. But clearly there will be a considerable data traffic that might be described as machine-to-machine communication. Elevators ordering spare parts or cars notifying mechanics of faults to be put right at their next service are two likely scenarios. NTT DoCoMo and Coca-Cola already have a joint project running to create I-mode-connected Coca-Cola dispensing machines.

Consensus is emerging on DoCoMo's assertion that cars will soon be using mobiles at a level that might be described as "revolutionary." Some luxury cars such as the Volvo 60 are already connected to the Internet. There is an automatic alarm if the airbags are inflated or the car stolen. In Japan, a research project is underway in which cars are being used as sensors for traffic intensity and weather. Data on the speed of windscreen

Table 5.2 Mobile customers year 2010 according to NTT DoCoMo

Mobile customers, 2010	Millions
Humans	120
Cars	100
Bicycles	60
Portable PCs	50
Motorcycles, toys, machines, pets	30
TOTAL	360

Source: Durlacher 1999

wipers, velocity and position of vehicles are fed back to the researchers, who are thus able to calculate weather and traffic conditions. Eventually, even more telematic solutions are expected in motorcars – such as satellite navigation, voice-activated mobiles, downloading and screening of films for children or passengers in the back seat, and so on. "Just as with airbags or ABS brakes, we believe that rudimentary versions of these systems will be standardized in cars within the next five to seven years," says Jan Hellåker, Managing Director of Wireless Car in Gothenburg, the company responsible for Volvo's On Call System.[37]

However, the costs of developing these kinds of systems are going to be high. Joint ventures between car manufacturers will become commonplace. In Europe, for instance, Ford is working with PSA Peugeot Citroen of France, which, in turn, is involved in joint projects with the media conglomerate Vivendi. Vivendi thus benefits from a complete value chain with everything from the content of car-focused portals, to intelligent features in the actual cars. Eventually, we may see the motorcar as nothing but an uninteresting steel shell. "Who's to say that Microsoft will not eventually buy Ford? We may actually finally come to see the car as nothing but the hardware, a small part of the whole. When in the future the customer buys software, the really important bit, he might buy it packaged in the form of a car," says the car manufacturing industry researcher Matts Carlsson at Chalmers University of Technology.[38]

Technical trends

Let's take a look at some other technical trends that hold significance for the development of the mobile marketplace, restricting our examination to technology that is tangible to the user or impacts on the success of the mobile marketplace as an advertising channel. Briefly and broadly speaking, we might conclude that the most significant aspects of developing mobile technology are:

- Faster data transfer capacity and expansion of broadband enabling "always-on" Internet connection.

- Improved and simplified equipment for use and presentation, a prerequisite for a mobile, connected life.

- Cheap and simple communications equipment making it possible to keep apparatus connected to the Internet.

Technology

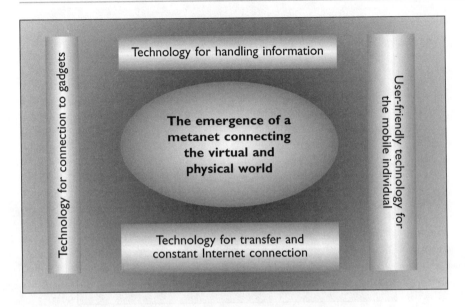

Figure 5.6 The emergence of a metanet

There are four principal technical dimensions of significance for the mobile marketplace as an advertising channel. These four dimensions lead to the emergence of a metanet that connects the physical and virtual world

- Improved technology for handling and working with consumer information – such as payments, and so on – enabling more tailor-made communication between companies and customers.
- Collectively, these prerequisites lead to the emergence of what might be called a metanet that connects the virtual and physical world, and allows individuals and machines to communicate in one and the same network (Figure 5.6).

User-friendly technology for the mobile individual

Briefly, there are four strong trends in the whole area of user-friendliness that will be instrumental in the creation of tomorrow's mobile technology.

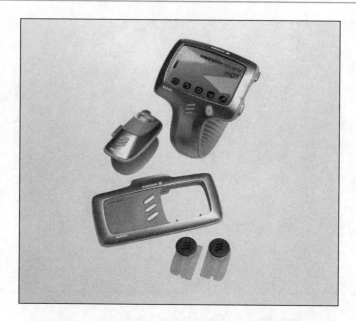

Figure 5.7 Miniaturization for user-friendliness
Miniaturization makes it possible to integrate computers or mobile transmitters in, for example, cufflinks or earrings
Source: Picture from Ericsson's future concepts
http://www.ericsson.com/press/phli_pcoco.shtml

Miniaturization

This makes it possible to integrate computers and mobile devices in, for instance, clothes, jewelry and spectacles (Figure 5.7).

Beyond printed text

There is no longer an explicit need for computers to communicate with the help of written text. New technology is proliferating in areas such as voice recognition and voice synthesis. It is already commonplace to have e-mails read to us by a digital voice via the mobile. And there is also software providing direct translation, so that people can communicate verbally in real-time across the boundaries of language. In the medium term, it is not inconceivable that systems will be developed to control computers with hand gestures. The overall effect of this movement away from text, will be more user-friendly systems.

The digitization of everything

This is the absolute foundation of the Internet and the mobile Internet. In practice, everything, even information disseminated via hard copy, is digitalized. Somewhere it is being stored on a hard drive in the form of ones and zeros, ready to be printed and distributed.

Communication is also being tailored to the needs of Internet portals, so that a universal standard is emerging for communication between computers, mobile devices, gadgets, and so on. The foundation is thus being laid down for a total exchange of information.

Increasing user-friendliness

After many years of intensive high technology research, the pendulum is now swinging towards a greater focus on user-friendliness. What this means, in practice, is better and simpler computer terminals and displays, clearer interfaces, better design, a smaller gap between product development and demand, as well as improved understanding of behavioral factors, needs and social patterns.

Tomorrow's mobile?

The mobile phone as it exists today is both a straightforward mobile phone and, occasionally, a telephone connected to a computer or palmtop device either by a cable or IR (infrared) technology. However, with the breakthrough of Bluetooth, there are already new possibilities of connecting (with great ease and without any wires) a host of different devices such as telephones/radio devices, handheld devices, headsets, or various interior or exterior devices such as monitors, keyboards and cameras, as shown in Figure 5.8.

The demands placed on these mobile devices are high, as well as hugely paradoxical. On the one hand they need to be small, light and energy-efficient. On the other hand they need to be large so that the screen is clear and the controls easy to use. In addition, mobile equipment is exposed to many more physical demands than stationary equipment, and hence needs to be tougher. And to make things even worse, conventional and established technology is usually more robust than the latest craze that hits the shops.

Energy consumption is one of the make-or-break questions. If devices are wirelessly connected, each one of them must have its own energy

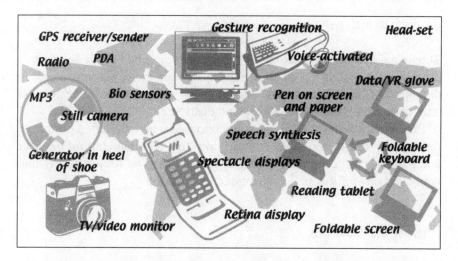

Figure 5.8 Mobile devices of the future
The mobile devices of the future will be used for a range of functions; they will be based on wireless connection between a wide range of devices

supply, whereas devices connected by cables can be powered from a single source. Available solutions to the problem of electricity consumption might be summarized in the following four ways: better, more lightweight batteries; fuel cells (fed with a sugar cube, a fuel cell might work for up to a week or more); generators (solar cells or generators built into shoe heels); new energy-efficient processors that adapt energy consumption according to usage levels. And new sources are needed. The batteries of the first generation of 3G-phones only last for 20 minutes of streaming video.

In the shorter term, the following accessories are also relevant:

- *Handheld computers.* There is already a range of available models, with or without keyboards, which can be connected via a mobile phone. During the course of 2001, an increasing number of these handheld computers will also be able to function as mobile phones.

- *Reading tablets.* A wafer-thin computer without a keyboard has long been on the list of desirables among those dreaming of digital newspapers and books. Commercial products including e-books have been available for a few years, but the reading tablet has not been a notable success. However, at the Comdex Exhibition 2000, the first prize for best "Vision for the Future" went to Microsoft's new Tablet PC (see Figure 5.9). Other companies are launching similar products. The reading tablet used in combination with wireless networks in the home, will probably

Technology

Figure 5.9 Microsoft's Tablet PC
This Tablet PC prototype developed by Microsoft demonstrates
the concept of tablet computing
Source: Microsoft

relocate the computer to the kitchen, sofa, bed and garden, and inculcate a whole new attitude to the way we use computers. Even so, it does not seem an entirely practical product to carry around in a city environment.

- *Spectacle frames.* Using glasses as monitors is something that has for long been playfully presented as an option for the future. It is now beginning to look like a realistic option. Various solutions can be envisaged – for instance, a screen in one of the lenses, or glasses with a foldable screen.

- *Digital speech.* Rather than receiving information in the form of text on a screen, it can be read out by a computer-generated voice. The technology has already been developed for disabled people and in call centers; mobile operators also offer a voice e-mail pick-up service.

- *Voice recognition* is a technology that has undergone significant improvement in recent years. Verbal exchange with computers is perhaps the most common interface in science fiction, but the question is how practical this would be in a real-life situation. General noise levels might become problematic and the atmosphere rather too chatty, if everyone at the office and on the bus were talking into their mobile terminals.
- *The electric crayon* is a technology already in use in devices such as the Palm Pilot and, before that, Apple's Newton. Its advantages are ease and flexibility of use. The main disadvantage is that handwriting recognition requires a fairly large sample of the user's handwriting style.
- *Screen Pen* is a new technology introduced early in 2000. The pen holds a small camera, and the paper a grid pattern, which allows the computer to know what is being written. The Swedish, Lund-based company Anoto was awarded second prize for this innovation at the Comdex Exhibition 2000, in the "Vision for the Future" category. Autumn 2001 Anoto, Ericsson and Nokia launched the chat pen CHA-30, based on existing technology in the Anoto screen pen.
- *The keyboard and mouse* have not been eclipsed yet. Small chat-boards can be bought for mobile phones, and foldable keyboards can be bought as optional extras with handheld devices such as Palm Pilot.
- *Radio receivers* for normal broadcast networks (radio networks).
- *MP3-players* for downloading music files from the Internet. MP3 is already available as an accessory to many mobile phones.
- *Digital cameras* for photographing images that can be distributed by e-mail to friends and family. Already available today in several mobile phones.
- *The GPS receiver* makes it possible to determine one's exact position. Already exists today in mobiles and wristwatches.

In the longer term, typical mobile equipment might be:

- *Digital paper and foldable screens.* Three large companies are looking for an electronic substitute for paper. Xerox, IBM and E-ink. E-ink, for instance, demonstrated a prototype of a 5-inch screen, which, in association with Philips, it hopes to be able to start delivering by 2003 – followed by paper-thin screens by 2005. IBM's concept is that electronic paper should as far as possible resemble cellulose paper, and has already indicated that a digital newspaper should consist of eight sheets. In the long term it is not inconceivable that there will be foldable or bendable screens strong enough to be shoved into bags or stuffed into pockets.

- *Retina display.* The image is projected directly onto the retina from an adjustable laser placed, for instance, in a pair of spectacle frames. The technology is not fully developed, but progress is being made. Sony has a pair of prototype video glasses, and a true retina display may be seen in laboratory conditions. If prices can be brought down to realistic levels, and if safety and comfort are acceptable, there is every possibility that retina displays could achieve a real breakthrough. Retina displays would give a larger surface area than even the biggest of conventional monitors.

- *Holographic monitors,* or monitors that in other ways create the illusion of a three-dimensional image may become significant in the context of mobile phone development. Many suggestions have been put forward, but as yet no wholly satisfying alternatives have been found.

- *One-handed keyboards* have been used in certain mobile computing prototypes developed at MIT. One key is used for each finger, and the different characters of the alphabet are produced by simultaneously pressing two or more keys at the same time. The technique is similar to playing chords on a piano keyboard.

- *The data-glove* is already being used in a variety of VR applications, and in the long term could be adapted for use in the context of mobile data communications. Designated gestures might be used to give commands to the computer, or a virtual keyboard developed similar to the one used to good effect by the hero in *Johnny Mnemonic.* In fact, virtual keyboards are already available on the market.

- *Wetware*, connecting computers directly to the nervous system, is probably the least likely of all these technical possibilities. Even so, there are certain precedents. Brainball, for example, is a game in which the player with the most relaxed brain wins. Another example already existing, is a technology used in certain virtual reality games in which a weak electrical impulse is passed through the brain, creating a feeling of movement. Films such as *The Matrix* and *Johnny Mnemonic* provide examples of how this kind of technology may end up being used.

- *Biosensors* are not as advanced as Wetware, being used merely to monitor biological factors. Scanning fingerprints or the iris are two ways of making identity checks. Sensors checking blood pressure and blood sugar levels are generally thought of in a medical context, but ultimately there may be other ways of using and connecting biological sensors to computers. What types of biological data could be used for the mobile services/products of the future?

Technological aspects of data transfer and "always-on" Internet connection

We have already alluded to the technical infrastructure that enables "always-on" Internet connections with high-speed data transfer capacity. In brief, the main issues are:

- *Improved data transfer speeds in wireless networks*. To a great extent, the main subject being discussed today is the transition to 3G networks. However, in practice there will probably be many different constellations of mobile telephone networks, local wireless networks in home environments as well as in the workplace and public places, and perhaps even broadcasting networks.

- *Data packaging in the networks*. Fixed-line telephone networks and even mobile telephone networks have up until now been circuit switched. What this means is that users are either connected or not connected. Packaging is simply a way of organizing information in convenient data packages. Users can remain permanently connected without incurring higher charges for information sent or received. Ultimately, messages will be exchanged much more quickly.

- *Expansion of fixed-line broadband* will create more opportunities for high-speed data transfer from computers. The evolutionary process towards broadband in the home makes it likely that users will eventually be demanding similar speeds in mobile equipment. Whatever is possible in stationary devices, will be expected also in mobile devices.

- *Expansion of digital television*. With the expansion of digital television, there will be further reinforcement of Internet-related habits based around the television.

Connected technology

We have already heard that machines are going to become mobile phone subscribers. DoCoMo, as we learnt earlier, believes that by the year 2010 it will have twice as many machine customers as human ones. This is something we might describe as the "smart-ification" of everyday life.

The process will be made possible by:

- *Better remote control systems.* It is already possible with available technology to use a mobile phone to call up a central exchange in the home, just to check that the oven is switched off, or to turn on some lights, the television, and so on.
- *Technology for local radio networks.*
- *Improved agent technology.* So-called intelligent agents are becoming increasingly influential, helping us choose products, gather information and monitor and carry out other tasks on our behalf. These agents can also to a growing extent act on their own initiative and show elements of artificial intelligence. How quickly they can learn to be really proficient we do not know as yet. There is also some uncertainty about how much trust we should place in intelligent agents.

The consequence of this is that there will be growing numbers of artificial mobile phone users, and a higher and higher total number of users. Machines will be customers, and they will be treated as any other flesh-and-blood customers.

From the opposite perspective – that is, that of the selling company – one might describe the product as an unpaid salesperson. At least that is the view of the m-commerce strategist and writer Pär Ström, who argues that Internet-connected products and appliances are set to bring about a new business revolution, precisely because they will function as online sales teams on behalf of their manufacturers.

Technology for handling information

To transform geographically- and personally-adapted information into tailor-made communication, a range of technical measures are necessary. In this area also, development is rapid.

- *Positional information.* Available technology has advanced in leaps and bounds.
- *Technology for filtering advertising messages.* Digital television, for example, will allow consumers to skip commercial breaks or specify product areas that are of interest. The technical foundation for this is PVR, Personal Video Recording. Dutch TiVo, backed by AOL and Philips, is already offering European consumers the possibility of storing up to 40 hours of footage for later screening. Eighty-eight percent of TiVo owners are currently skipping commercial breaks.
- *Payment and identification methods.* If the mobile marketplace is going to be a success, secure identification procedures must be achieved. There are many available technical options.
- *Technology for making minor payments via the mobile phone.* This technology has been in use for several years, for instance in Finland.

The crunch question in the long term is whether it will be possible to personalize market communications at a reasonable cost. Will it be possible to automate the selection of information that goes out to users? On this point there is a considerable clash of expert opinion.

DIGRESSION
Per Florén

What happened (and is happening) to the software agents?

For a few years in the mid-1990s, so-called "intelligent agents" were a popular theme in most visions of the future of the Internet. Agents would venture out there into the virtual world, collecting news, sorting information, tipping us off about good television programs, finding cheap holiday offers and tracking down the best prices on the kinds of goods and services that we, their owners, wanted to buy.

Not only would these little agents search the *entire* Internet (this was the phrase used by many of these Internet visionaries), but they would also learn our habits and preferences with such a degree of accuracy that after a while, CDs and air tickets would more or less drop through the letter box automatically.

A good example of one of these early, prognosticating articles published in *Fortune* (24.1.1994), was "Software Agents will Make Life Easy."

Many companies were established to realize these visions. Among the household names were General Magic, set up by a group of defectors from Apple in the early 1990s. The research company Ovum predicted in a report in 1994[39] that agents for message processing, information retrieval and customer-adapted solutions would be turning over almost US$3 billion by the year 2000. Experience did not bear out their predictions.

Today's expectations are much more moderate. General Magic has turned its attention to voice-recognition, and the true intelligent agent of the future is still conspicuous by its absence. Some research is still going on, in Sweden and elsewhere, but there is an enormous gap between what is happening now and the ebullient vision of the 1990s. Meanwhile, research is going ahead at institutions such as SICS at Kista (Stockholm), Sweden, into how agents should be constructed in the future, how they should function, and which particular services lend themselves to particular agents.[40] Perhaps there will be a breakthrough sooner than we think?

Why did the agents not live up to our expectations?

It appears that there were basically two reasons for the failure of the program agents.

First, it proved too difficult to create the level of intelligence required to satisfy even the most minimal of user demands. Artificial intelligence, AI, has always been further into the future than many are prepared to admit, and this is particularly true of software agents on the net. The inadequate intelligence of the agents makes it difficult for them to understand and evaluate information they find on the net, and ever harder to relate it to the needs of their owners/taskmasters. To put it bluntly, they are simply too dumb to do their jobs. Small wonder that most of them are out of work!

Second, the intelligent or half-intelligent agents that were built did not survive in the tough and competitive virtual ecosystem of the Internet. In ecology, the term "evolutionary race" is used to describe the competition that is constantly going on in the biological world between predators and their prey, between parasites and the host animals, between viruses and humans. Every improvement in the predator will almost immediately produce a commensurate improvement in its prey. If the cheetah learns to run a little faster, the gazelle will quickly learn to run a little faster too. If for instance the aim of intelligent agents on the net were to identify the cheapest price for a particular book, then the evolutionary race for all book retailers would be how most effectively to stop or lead the intelligent agents astray. In this race, the retailers have up until now always come out on top. It has proved easier to find ways of stopping the agents than to create smarter agents.[41]

The trends of today

Intelligent agents seem to have developed in three directions:

- More intelligent search engines that help the user to find a particular piece of information or service on the Internet.[42]

- Relatively limited intelligent functions, for instance network programs that monitor how well the Internet is working, and take certain prearranged measures to prevent or remedy various problems.[43]

- Highly structured systems for communication between services on the Internet, such as Hewlett-Packard's e-speak. The system is based on a defined way for the featured services to respond to questions, negotiation, and so on. There are also a number of registers storing information about which companies can be accessed on the net. This type of system will probably work primarily with groups of companies whose price sensitivity is not high.

A prognosis?

If we proceed on the basis of what we have today, and the lessons learnt from the experience of the last decade, a realistic prognosis might be:

- Search engines will slowly improve, become faster and better at locating just the kind of information that the user requires. No fundamental developments will happen in the next five years. However, mobile users will experience step-by-step improvements in portal and search functions – these will find and present information and products that the user wants.

- Industrial standards for the exchange of information on the Internet, such as Hewlett-Packard's e-speak, will slowly be established over the next five-year period. However, their usefulness will be limited while only a limited number of services are connected to the service. In the current climate it is difficult to predict what type of industrial standard will become dominant.

- Methods of hindering search engines and halting or confusing price comparisons will keep developing as quickly as the agents. This "evolutionary race" will carry on for a long time, perhaps as long as the Internet remains. The only way is for agents and vendors to find a win–win relationship.

For the mobile marketplace, the implications of this are:

- Agents will only be a viable option in relation to retailers that want to cooperate, perhaps in some kind of virtual mall. Companies that wish to take part must provide the required information in a certain standardized way. However, agents that are smarter than the information sources are probably a long way off, perhaps an infinitely long way off.

- The standardization of search methods, such as for instance the price of a book with a particular ISBN number, would be a useful addition and probably already is so. Certainly, standardized search methods would help someone in a shop, wanting to find out whether a price was competitive or not.

- Agents monitoring certain types of news or offers might soon be a working reality, but only within limited frameworks. "Do you want to buy a cheap bucket seat? Enter your name here, and you will receive an SMS as soon as there's a spare seat." Or "Do you wish to buy shares in a certain company, but only once shares dip below US$35? An e-mail read out over the phone will give you the required information."

- For more complex and universal program agents, however, there will still be a delay of several years, perhaps even decades.

... A final thought

If truly intelligent agents really did arrive in a few decades, would this benefit people? Would it be a good thing for us to hand over choices – what news we read, what e-mails we are prepared to read, what airlines we fly with – to the software in our computers? Would we not be weakening our own capacities to sort information and make choices, in the same way that young people today are less able than in previous generations to do mental arithmetic?[44] Will agents one day find that humans are rather a disappointment? Will they conclude that we are not living up to their expectations and take over the world?[45]

Technology

SUMMARY

- The pace of technical development and network utility is exponential.
- New technology is introduced at an ever-quickening pace.
- Mobile telephone networks, wireless local networks and Bluetooth will lay down the foundation for the mobile Internet and constant Internet connection.
- Entertainment accounts for half of all the services in the Japanese I-mode.
- Cars, refrigerators and dogs are going to become mobile phone subscribers.
- The mobile headset will become a central hub wirelessly connected to and communicating with a range of other gadgets.

CHAPTER 6
The Institutions

Social, economic and legal incentives and vehicles for new technology play a major part in its eventual breakthrough. A lack of appropriate rules or vehicles can significantly inhibit technical development and the eventual breakthrough of new technology. Ultimately, what this is all about is institutions.

In this chapter, we look at the institutional prerequisites for the development of a mobile marketplace.

The importance of institutions

First, let's look at a few historical examples. The industrial revolution in England at the end of the 1700s was not caused primarily by the technological state of the country. Nor was it stimulated by a better education system. France and Spain were no less preeminent in these fields. Rather, it was a case of the institutional preconditions being firmly in place: private ownership was well established by law, and an entrepreneurial culture was accepted and even seen as something worth striving for.

If we observe successive waves of industrial development between the late-18th and the early parts of the 20th century, each long-term economic upturn was clearly based on a cluster of new technologies carried forward by new types of institutions – from the smallest family-owned company to transnational behemoths. Institutions create preconditions for the breakthrough of new technology, while at the same time themselves being subject to the shaping forces of those technical advances.

Government support for technical development needs to be firmly included as one of the "institutional preconditions." The almost symbiotic relationship between Ericsson and Telia (then the state-owned telephone network) established the foundations for the subsequent technical development in a company that has become one of the acknowledged world leaders in its field. The same might be said about the relationship between SJ (Sweden's state-owned railways) and the company formerly known as Asea – now a part of ABB. In both of these cases, state-owned companies functioned as vehicles for new technical systems, and state investment leveraged structural change on a broad front.

However, while state involvement can act as an incentive, it can also inhibit development. French investment in Minitel may have been successful, but ultimately it put France at a disadvantage once the Internet race started in earnest. Critics of the digital, terrestrial television network maintain that the same situation will arise in the television sector, in countries now investing in terrestrial, digital television.

Regulations and monopolies can also hamper the development of new technical systems, as illustrated by the history of the telegraph. When Samuel Morse developed the telegraph in the early 1830s, its technology effectively changed the world in a matter of half a century. The telegraph made it possible for companies to grow into transnational organizations, which led to a decline in the strong position of merchants and import–export companies. However, the real problem of the telegraph was that it was complicated and expensive; in addition, the networks were often controlled by state monopolies. As a result, it never developed into a consumer product.

The telephone, on the other hand – even though initially developed as a by-product – quickly established itself after Alexander Bell's patent application in 1876. In its early days it was viewed as more of a toy than anything else. In Stockholm, the first private lines were already installed by 1877. The networks grew quickly and extensively, unhindered by the sorts of regulations that had held back the telegraph.

With reference to the mobile marketplace, institutional prerequisites can be broken down into two main components: First, macroeconomic change (that is, in companies) and new patterns of work; and second, institutions and legislation.

The emergence of the knowledge-industrial society

The transition to a knowledge-based industrial society has been, and will be, the main driving force behind the emergence of the mobile marketplace. Impetus for the process will come from technical development, deregulation and a legislative level playing field. These factors will eventually create an entirely new business logic for the active players in the mobile marketplace.

The knowledge-industrial society is the organizational and legislative outcome of a range of structural changes. It will ultimately emerge out of a combination of deregulation, globalization of economies and manufacturing systems, technical development and communications.

The result will be a social system affected by:

- Globalization and the 24-hour economy.
- Hyper-competition and a frenzied hunt for customers.
- Automated production of goods and services.
- Shorter product lifecycles and "Biotopia."
- Joint ventures in development and markets.
- A move towards knowledge acquisition and knowledge organizations.

The post-industrial dilemma

The post-industrial dilemma is both one of the motors of the knowledge-industrial society, and one of the results of it. In broad terms the post-industrial dilemma is about the rising cost of manpower in relation to the cost of the product – a consequence of the fact that we are manufacturing goods ever more quickly and cheaply (see Figure 6.1). As a result, automation is being introduced in every conceivable service sector. Cash dispensers and Internet banks are replacing banking clerks, and telephone operators are being usurped by computerized telephone exchanges.

The post-industrial dilemma is fertile ground for the mobile marketplace. Just as with the Internet, customers are being invited to take part in (and create) the value chain, primarily for services but also products. Additional services can be set up once mobility and positioning have been implemented and integrated into the framework of the virtual marketplace.

However, the post-industrial dilemma is also one of the most serious challenges for the mobile marketplace, from a communications or advertising perspective. Only if tailor-made advertising (specific to time, place and identity) can be automated – something that the advertising industry is pinning its hopes on – will the mobile advertising market ever become a reality.

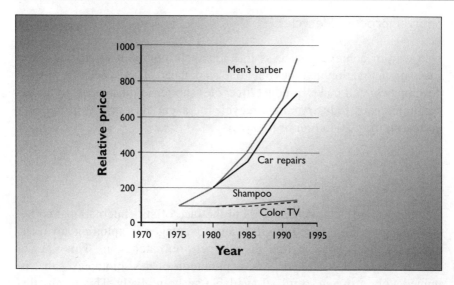

Figure 6.1 The price of a haircut in relation to shampoo over a 20-year period

Hyper-competition

Hyper-competition is a result of maturing markets, globalization, overproduction and declining sectors. Eventually it leads to an increasing difficulty for companies to identify long-term, sustainable competitive advantages. To succeed, they either have to grow or get narrow – get big, get niched, or get out!

Dismiddlesizing

One of the consequences of hyper-competition is that it becomes more and more difficult for a medium-sized company to survive. This is known as "dismiddlesizing" – that is, middle-sized companies are disappearing. The survivors are either large and cost-effective, or small and extremely specialized. There is also a smattering of small, local companies. The world thus becomes analogous to a rainforest: with rich undergrowth, a more or less nonexistent middle layer and a gigantic canopy of leaves at the top.

Shorter product lifecycles

One way of achieving success in intensively competitive markets is to focus on innovation and thus stay one step ahead. This speeds up product development and makes product lifecycles shorter in most areas. In the car manufacturing industry, the lifecycle of a car has in recent decades decreased from 20 to four or five years.

Biotopia

Another way of achieving success in the face of stiff international competition is to rapidly make full use of any advantages – exploiting any niche that has been created, before the competition catches up. This kind of activity might be described as an exploitation of new biotopias – that is, immediately growing to fill all available space globally. This means that even small, newly established companies are looking for global presence right from the start, with varying degrees of success.

"Finishing second or third is not enough on the global stage if the total market consists of no more than 15 000 end users" commented Hans Johansson, CEO of the biotechnology company Personal Chemistry.[46]

From the material to the weightless economy

Not so long ago, if we wanted to have cake we would buy flour, sugar, eggs, baking powder, cream and strawberries. Then we would bake the cake! More recently we started buying ready-made sponge cakes or ready-to-eat cakes. Parents now increasingly buy birthday parties, with the cost of the raw materials a mere fraction of the whole; the transaction is about buying an experience, whether the party is held at McDonald's, Hard Rock Café, or a theme park.

This process traces the movement away from a material towards an immaterial economy. We are no longer paying for anything that can be quantified by weight or size. Rather, we are buying an experience, an emotion – something that makes us feel like better people.

What we are buying might be brands (and the emotions conveyed by these), or tangible experiences such as a camping trip in a nature reserve. All of these transactions have one thing in common, namely that we value something indefinable, something weightless.

The transition to an immaterial economy means that a greater share of the overall value is created in the development chain (new products with short lifecycles, new software, and so on) and in the marketing chain (the positioning of "the BMW feeling," and so on). Less and less value is added in the production. Products, insofar as they exist at all, are expected to be perfect. At any rate, when production costs are practically at zero levels (such as online delivery of digital products), the scale advantages are enormous.

From a transaction to a concept economy

The transition to an immaterial economy also means that we increasingly buy not products but concepts. Car owners do not simply buy a car. They also buy a dealer network, a servicing organization, a group of car-owners (to which group they will henceforth belong), second-hand value, and so on. Company purchases of products and services from subcontractors are similarly affected: purchases are increasingly whole systems and functions rather than specific services or components.

Joint ventures in the development chains

With more and more value created in the development chains and shortening product life cycles, development costs are on the rise – and hence the sales volumes required to justify these development costs are also growing. In the car industry, joint ventures and mergers have been in evidence for at least a decade; including joint ventures in R&D. Many car models are also being built on the same platforms. The VW group has four major brands: VW, Audi, Seat and Skoda. But all the models are built on a small number of shared platforms.

A similar trend is noticeable in the telecommunications industry – for instance, in Toshiba and Siemens' earlier joint venture in 3G mobile telephony. Ericsson's collaboration with Sony, and Philips' joint venture with China Electronics illustrate the same phenomenon. The underlying reasons for a parallel wave of mergers in the pharmaceutical industry are much the same.

Collaborative ventures in marketing

Just as collaborations are becoming more common in product development, they are also becoming widespread in marketing. The reason, of course, is that the customer has become the narrow sector. The costs of gaining new customers have increased, and companies have to make the process as cost-effective as possible. Besides, consumers group themselves on the basis of lifestyle and values – hence it may be advantageous for companies to position themselves close to other companies with similar profiles and target groups.

Co-branding and co-positioning are recent coinages, which cover aspects of this phenomenon. The alliance between McDonald's, Disney and Coca-Cola is often viewed as the most successful, high profile example. Alliances based on loyalty cards and bonus systems, not least for airlines, are other instances of collaborative marketing. In November 2000, Scandinavian Airline Systems launched its concept "SAS Marketing Partnership" with the express idea of helping "brands with similar values strengthen their positions and find synergies."[47]

Owning the customer relationship and share of mind

Anyone wanting to keep hold of the customer, should aim to get as large a "share of mind" as possible, as well as some sort of financial relationship. Because the memory is dynamic and tends to start eroding if actions are not reinforced, it is crucial for communicators to maintain a more or less daily presence in the consumer's mind. In the bank/insurance equation, the bank is dominant. Contacts between the bank and the consumer probably occur at least once per month. However, statements from the insurance company are unlikely to be forthcoming more than once per year. To create a continuous customer relationship, many companies and financial institutions are now establishing portals and interactive websites. This is one way of prolonging the dialogue, to keep the brand alive, and carve out a "top of mind" existence. Pure portals, on the other hand, have a precarious existence, as they usually lack a financial component that gives rise to revenue streams.

Stock exchange leadership, gorilla logic

More and more structural change in the global economy is steered by stock exchange valuations. Share values are the currency. This has been patently clear in the Internet business over the last few years. Successful companies achieve higher valuations than less successful ones, and thus acquire strategic advantages. By using their own shares as currency, they can buy competitors many times larger than themselves, in the hope that the acquired company will afterwards be valued at similarly high levels. The collective value will thus have shot up substantially, and the purchasing power of the new company will have grown by a greater multiple than its actual growth in size. In this way, strong and high-valuation companies (known as "gorillas") can buy up their competition.

This is most obvious when different yardsticks are used to value companies. In what was formerly called the "old economy," companies were primarily valued in relation to profitability. In the "new economy," companies were capitalized on the basis of growth, or turnover. What this actually meant, was that a small company capitalized by the "new economy" yardstick could take over a much larger company valued by traditional criteria. An excellent example of this was a bid for the London Stock Exchange made by the OM Group at the end of 2000. The OM Group had achieved a high valuation as a result of its investment in new transaction systems.

Innovation in small companies

As a result of various valuation principles, new technology is increasingly being developed in small companies. In a traditional company whose value is based on its profits, technical development is a drain on financial results. In a small company, where some of the development costs might be covered by joint ventures or kept to a minimum by initially running pilot projects, technical development increases turnover. If the company is listed, such projects will drive up valuations.

For large, traditional companies it is thus advantageous to buy fully developed technology or patents, even if the costs are higher than would have been the case if developed internally. Buying technology saves time, and also avoids the risk of running up additional costs that might have dampened stock market valuations. Acquiring companies is also a way to corporate renewal.

Patents and technology companies have been purchased on a large scale in recent years, not least within the pharmaceutical industry and the IT sector. Cisco is one of the companies that have used this strategy with great success.

The movement towards knowledge-based work and knowledge organizations

With the transfer of value creation towards development, marketing and selling, more and more companies take on the characteristics of knowledge organizations, and more and more employees become knowledge workers. In practice, this means that personnel are less dependent on a physical workplace, but more dependent on a meeting place.

Work is thus liberated from the limitation of space. Ever larger numbers of people have to do a larger share of their work away from the workplace, and a practical mobile support makes this a much more feasible proposition.

The evolution towards knowledge organizations also brings about a convergence between work and leisure. If people do their work at home, their private lives begin to coexist with their professional lives. Living an organized life is just as much about having a tidy personal life as it is about professionalism.

Rules of play, legislation and state incentives

We now briefly touch on a few central questions regarding the rules of play in the mobile marketplace, including: differences in standards and players in 3G networks globally, the struggle to establish universal legislation, deregulation, copyright issues and privacy.

Different players and standards create different preconditions

The future of the mobile marketplace rests on strong vehicles for technical systems combining with potential users and their preferences. There are marked differences in different parts of the world. In Europe there is a well-established GSM standard, stable network operators and high levels of mobile phone use. In Northern Europe, furthermore, Internet use is high; while in Southern Europe, people have not really "taken the plunge" into the Internet habit. Hence in Europe there are relatively good preconditions for a fast-track expansion of the 3G networks and mobile marketplace, perhaps especially in Southern Europe where Internet usage has not yet taken off.

In Japan, the situation is even more promising. I-mode is a practical if embryonic version of a fully working mobile marketplace. Assuming there is a rapid expansion of 3G networks, and assuming that the system works with sufficient speed, there are good prospects that Japan will leapfrog the Internet and go straight to the mobile marketplace.

In the United States, the situation is different; and the difficulties of establishing a mobile marketplace are far more entrenched. One of the obstacles is that, for legal reasons, there is a very inefficient mobile phone market – one in which the receiver pays for the call. As a consequence of this, many Americans would rather turn off their phone – that is, the few of them that actually own one – rather than risk receiving an expensive call that they do not even want. In addition, the market is split into several technical systems, making it difficult to phone from one region to another. A final hindrance to the emergence of a fully fledged US mobile marketplace is the high level of Internet usage, and the fact that users have become accustomed to free content.

The quest for universal rules and increased trade within the EU

E-commerce is not affected by national boundaries, and hence fits comfortably with the principle of the European single market. Creating a single market has, in practice, not been an easy matter. The EU has shown enthusiasm for e-commerce as a fledgling market and has exerted itself to stimulate the growth of both e-commerce and mobile commerce. An action plan known as *eEurope 2002*[48] has been produced to outline the vision of the information society.

The biggest remaining challenges for B2C commerce are, first, the remaining legal variations, which cause problems for SMEs (small and medium-sized enterprises) involved in commercial activity across Europe. Second, there is poor consumer confidence in the idea of making payments across the Internet. Establishing the new domain name .eu is another way of stimulating commerce right across the Union.

Measures that have been cleared for action include new copyright legislation, e-money and new quality kite marks to build consumer confidence. Brussels expects to complete most of the new legislation within the next few years.

The EU has also looked at the whole issue of taxing goods and services purchased via the Internet. Two main principles have been established. First, no new taxes will be levied without first adapting the existing system to harmonize with e-commerce. Another principle is that products (categorized and taxed as services) will be taxed in the land/region where they are consumed.

Both at national and international level there are clear moves to regulate e-commerce. Rules are often seen as something negative, but success in the mobile and e-commerce fields will depend on consumer confidence – and proper regulation should play its part in the framework that eventually builds consumer confidence.

Increasing deregulation

There is a huge wave of deregulation sweeping the world, replacing state monopolies with competitive markets. The development may be seen in telecommunications, banking, finance, insurance, energy and pensions. The repercussions are not only significant for the structure of the global economy, but also for individuals and how they behave.

Deregulation opens up markets that were previously inaccessible to competitors, forcing national players to go out into the international market. This development feeds into what was earlier called hyper-competition. In the long term, sectors previously dominated by national companies will be taken over by a handful of global behemoths. Airlines are heading this way – after a wave of mergers and the 2001 airline crises – and telecommunications and banking seem to be moving in the same direction.

The development impacts on consumers. Basically, it stimulates the creation of the Homo Consumentus by forcing people to make active decisions about electricity suppliers, telecommunications networks, schools and pension schemes. Europe, in its earlier guise, was a place where most of us did not have to consider such things.

Copyright becomes a critical issue

When printing was invented it suddenly became possible to copy and publish books, and with this, copyright law came into existence. Since then, an extensive system of rules has been built up to protect the intellectual rights of creators to their own works.

Fundamentally, there are three significant changes that make the whole issue of copyright a hot issue at this time: digitization, globalization and the increasing value of immaterial products.

While artistic works were purely physical, in the form of photographs, books, 78 rpm records and LPs, copying was relatively complicated and mass distribution of contraband fairly costly. However, the digitization of everything has made copying and mass distribution infinitely easier.

If digitization makes it easier and cheaper to make copies and mass-distribute them, globalization and the growth of "immaterial" values have made the consequences of this piracy much more far-reaching. The globalization of culture and media has created potentially much larger markets for creative works, and thus higher potential losses for the holder(s) of the copyright. If the Chinese are listening to the same music as Westerners, financial losses will be devastating if the Chinese Government does not enforce copyright laws and looks the other way as pirated copies flood the market. A uniform copyright law in all countries is therefore (according to many experts) essential to the creation of a global, digital entertainment and information society.

Many analysts are doubtful about the feasibility of a foolproof legal and technical system for protecting intellectual property rights, even if in principle all countries can reach agreement on the guidelines of such a system. Three problem areas seem to suggest that the long arm of the law will always fall short:

- *Perfect copies:* digital copying makes it possible to achieve perfect copies very quickly and without any appreciable reduction in quality.
- *Anonymity:* the consumer/user is nowadays often anonymous, or at any rate very difficult to trace.
- *Fragmentation:* before, a creative work was bound by the physical limitations of the medium. Now, parts of the creative work can be dispersed in new ways.

These three factors as a whole mean that copyright law is difficult to enforce. The problems caused by countries not adhering to international agreements – openly selling pirated music and software – are dwarfed by the larger issue of digitization.

The problems of copyright are clearly exemplified by the trial in the United States against Napster; or, indeed, the case of a trial brought in Sweden against a user who had posted links on his homepage to MP3 files – eventually, a verdict was returned that no law had been broken.

Even if in the final analysis it proves impossible to create totally watertight copyright protection, it should be possible to set up a system that offers a measure of protection vis-à-vis the creators of original material. This has, for instance, been achieved in the software market in spite of the fact that illegal copying of both utilities and games is rife. Furthermore, if MP3 files were available at reasonable cost, the vast majority of listeners out there might actually be prepared to pay something for their favorite songs. If one or two of them insist on getting a free ride, it might not be possible to prevent them every time they try.

Confidentiality – an increasingly hot issue

Whenever people surf on the Internet they leave electronic footprints, that is, details of sites they have visited and actions carried out. Many websites collect much more information than this, demanding, for instance, that users register before being allowed access to their sites. Member registers are an extremely valuable resource; and information about people's interests and preferences is often in real terms a sort of payment for services that are presented as "free" in a monetary sense.

It seems increasingly likely that the question of personal integrity and privacy will be the make-or-break issue of the computerized society. As consumers, we give careful thought to any disclosure of our personal details. The mobile marketplace is likely to be affected by a similar reluctance. When do we actually want to be reached? And when do we get an unpleasant feeling that we are being watched, like characters from George Orwell's famous novel, *1984*?

Although privacy is a fundamental human right, it is countered by a desire common to all people to be recognized – and to "be someone." The balance between these two positions is difficult to establish.

Increasing exposure

Communication via the Internet and other network services exposes people to scrutiny. Electronic footprints can be used in a number of ways, most of which are not understood by the general public. For instance, information assembled from all sorts of unlikely sources can help build profiles that indicate how people might behave under a different set of circumstances. In the IT world this is known as *data-mining*. It has been claimed that credit card companies know in advance when a divorce is about to happen, simply by looking at the transactions of the husband and wife – it's a safe bet that both will be withdrawing as much cash as possible. People buy more paints and brushes and white goods immediately after moving into a new house. By tailoring one's marketing effort to their needs, much better returns can be achieved per unit of advertising expenditure. With the introduction of position information in the mobile marketplace, this sort of thinking will quickly become the norm.

This development is neither intrinsically good nor bad, merely a consequence of technical evolution. On the face of it, it would seem quite useful to be sent a message about a special offer – a cut-price item that one actually wants to buy. Is it necessarily intrusive to have one's electronic footprints stored and later used to ensure more such personalized messages?

Privacy is weighed against the usefulness of information or advertising messages. However, uncertainty about what kinds of electronic footprints we leave when we use the Internet makes it impossible in principle for most people to evaluate how important the loss of privacy actually is.

Project Echelon – our "Big Brother"?

In the film *Enemy of the State*, Will Smith wages an unequal war against the US state security mechanism. Many people who see the film ask themselves if this could ever happen in the real world. Could the state really control its citizens to such an extent?

Others maintain that on this occasion fact exceeds fiction by a good margin, and that all Internet traffic is already being watched by the monitoring system known as Echelon.

Echelon is supposedly a global listening system with the intelligence services of the USA at its hub – and further participation from the UK, Canada, Australia and New Zealand. Advanced search programs that raise the alarm whenever key words are recognized are allegedly scanning a large part of the world's data and telecommunication traffic. According to Echelon Watch, a division of the US civil rights organization ACLU, sources have claimed that up to 90 percent of all Internet traffic is being monitored. The information is used partly for industrial espionage and partly to combat drug smuggling, terrorism, and so on.

Even though Echelon has been mentioned in the media, the debate has been surprisingly low key. Perhaps when all is said and done people do not want to be reminded of the fact that they are being watched and monitored. Many countries have set up their own investigations into the matter. Certainly, some of the silence about Echelon is caused by the "no comment" policy of the governments and security services that allegedly set it up. Most of the claims about Echelon have not been commented upon at all; and the only consistent government denial is that any illegal monitoring of citizens is going on.

Even if the rumors about Echelon are not all true, the ability of hackers to get into our computers and read our e-mails is not in doubt. The private domain is shrinking and people not only *feel* watched, they actually *are* more watched.

SUMMARY

- Institutions in the form of culture, rules and organizational forms create prerequisites, development and breakthroughs for new technology. New technology also creates new institutions.
- The knowledge-industrial society, marked by hyper-competition, knowledge, information logic and movement towards the periphery of the value chain (development and marketing), is the framework inside which the mobile marketplace begins to grow.
- Differences in existing players and rules in different countries, efforts to create a level playing field, deregulation, copyright issues and the privacy question, are factors that will impact on the future of the mobile marketplace.

CHAPTER 7
The Individual

Either get rid of technology or accept its acceleration. (Gunnar Thörnqvist)[49]

The speed at which we live life is definitely increasing. At the same time, we no longer have to bake our own bread, wash our clothes by hand or slaughter the pig. One of Microsoft's advertisements says, "Do things quickly, leaving more time for fun." The truth of the matter is that a lot of supposedly timesaving devices end up taking all of our time. One of the most important prerequisites for a broad acceptance of mobile habits and services is the relationship between technology and individuals. The heart of the matter is summarized in the quotation at the top of the page. If we are going to live with technology we have to accept acceleration in our pace of life. But whatever popular wisdom might be, it is people and not technology that control the process, at least according to Britt Östlund, Head of Department, Use and Effect of IT at KFB (Swedish Agency for Innovative Systems).[50] "Technology is not deterministic. Producers tend to believe that it is all about information, and that the users have not fully understood what the technology is for. But technology is malleable, technology is soft – it is *people* who are hard ... ." Phasing out technology is no longer a possibility, hardly even desirable. Meanwhile, the quickening pace also brings a quickening rate of change, so that knowledge and values are outdated much more rapidly.

If the technical preconditions affect what is technically possible, then the institutional preconditions affect organizations and business logic (including, to some extent, the types of services that might do well in the marketplace). Once we get on to the human preconditions, we have to look at the actual mobile marketplace, and ask ourselves what kinds of services are actually going to be in demand. How will people want to use the technical possibilities? Let us begin by looking at a couple of the major people requirements followed by an analysis of some clear consumer trends that are likely to impact on the mobile marketplace.

First, let us emphasize that the major prerequisites from a people perspective are based on five heavy trends with roots reaching right back across the centuries – although in recent decades they have become if possible even clearer:

- Individualization and personalization.
- Postmodernization, pluralism and global tribal societies.
- A revolution in diversity, meritocracy and higher ambition levels, and a need to reduce decision-making.
- Time-effectiveness, contemporaneousness, relevance and convenience.
- The increasing professionalism of the consumer.

Postmodern values – the victory of pluralism?

We live in an age of transition between old and new – the modern and the postmodern. Among the most significant shifts in value systems is the transfer from the collective to the individual. Increasing individualization establishes a greater emphasis on personal preferences irrespective of social factors. Earlier boundaries between class, nation, race and culture have been superannuated and are diminishing in importance. Globalization is bringing a whole new global culture, a so-called globally homogenous lifestyle. We tend to all wear the same designer clothes, watch the same television programs and listen to the same music whether we live in Stockholm, New York or Kuala Lumpur. In parallel, there is much greater heterogeneousness at local level. Individuals from different cultures and lifestyles might be living next door to each other, but their true loyalties are with cultures and lifestyles on the other side of the world. This diversity obviously contributes to further changes in ethics and values. Another crucial and entirely new value change is the game of identity and lifestyle that is now widespread. Globalization has opened up the world and the ability to communicate and gain impressions across all boundaries is infinite. What sets people apart from other people nowadays is their consciousness, their sense of having their own thoughts, values and lifestyles. There is certainly a strong sense of being able to choose one's own identity, and then express it by means of various exterior attributes. Clothes, cars, where we live and work, and what kind of mobile phone we have – all these things say something about us. By and large, we enact our personality by means of symbols and attributes loaded with values by the advertising industry.

What will we want to consume?

One of the main driving forces of the mobile marketplace is the desire to gain time – lack of time is one of the main problems that people experience in the 21st century. The desire for time-efficiency prepares the way for utility services, services that make our day-to-day lives a little simpler. At the same time it seems unlikely that banking services are going to be as important as has been predicted; the reason being that people are accessible to certain kinds of services while mobile, and to others services while static.

The mobile marketplace has to offer added value if people are going to start using it. "What is available on the Internet or the mobile Internet must be better than the alternatives already out there," says Solveig Wikström, Professor of Economics at Stockholm University.[51] "Technology might produce a short-term craze but in the long term its fate is completely decided by what it actually offers in terms of utility," she continues.

The real winner looks like being the communication channels. Earlier we spoke of "empty channels" in which people create the content, such as e-mail, SMS and telephony. Communication, like creativity, is a human necessity and prerequisite for living. Music services are growing fast on the Internet, presumably because the user can download individual preferences and create his or her own compilation CDs. This is another instance of an empty channel filled with content that we have actually chosen ourselves.

The importance of the brand

> Prices do not rule the web, trust does. (Frederick F Reichheld and Phil Schefter)[52]

In the mobile marketplace the buyer and seller are separated, perhaps both by time and place. The distances between the various players in the configuration are much larger than before, and hence the brand becomes ever more significant. Well-known brands inspire confidence, and thus it becomes more acceptable to make a transaction. In addition, if customers have confidence in a supplier they are far more likely to divulge personal information.

At the same time we have a tendency to be disloyal to brands because of the multifarious identities and lifestyles we assume. Small, unknown companies have very poor prospects of growing in the virtual marketplace; there is simply no consumer confidence out of which to grow. There might be the odd exception of a company that is first in a new niche, and thus has a head start. However, it is not only confidence that plays a crucial part – there is also the matter of values attached to the brand, significant because of the fact that consumption is a lifestyle.

As a consumer, I buy a set of attributes that fit with my conception of what I am today. My loyalty is thus entirely conditional, I will only stay in this place for as long as it suits me – when I am ready, I will break camp like a nomad and head off towards the next oasis.

Even the strongest, most well-established brands are aware of this development: very few consumers are consumers for life. As a player in the mobile marketplace one has to constantly recapture the consumer. In Nokia's commercial for the 8210 model, the actor says: "I'm always on the search. So far I've found more than I was looking for." Nokia and other companies are hoping that not only this customer but also all of his friends will keep finding exactly what they want with them. But, as the customers themselves are making clear, they are always on the lookout.

Open landscapes or the tyranny of diversity?

What do modern individuals, with their nomadic tendencies, really want? Open or closed landscapes? Enthusiasts like to allude to Metcalfe's law on the exponential value growth (n_) of networks with rising numbers of users. The great impact of the Internet can certainly be partly explained by this phenomenon. The network increases in value the more people that take part in it. But there is also something that might be known as closed landscapes – something that might be dubbed "limitations in mental bandwidth."

Kevin Kelly, one of the pioneering thinkers in the emergence of the New Economy, writes in his book *New Rules for the New Economy*: "… Because communication – which in the end is what the digital technology and media are all about – is not just a sector of the economy. Communication is the economy."[53]

But communications is not just economics. Communication is also the foundation for our society and the basis of our identity. Every obstacle to free communication, with whomever we want, whenever we want, is ultimately going to disappear. "The new thing is that we can communicate or buy services while we are moving about," says Magnus Bergqvist, ethnologist at Viktoriainstitutet.[54]

Network operators, mobile phone manufacturers and content providers are building portals today so that they can keep hold of the users, increase the click-through and make money by offering services. It is a sender-driven development, but there are some clear indications that the open landscapes are starting to close down, at least initially – this trend might be described as a threefold development towards reduced decision-making, and increased convenience and privacy.

The availability of services will be enormous, and all we have to do is choose. The mobile marketplace needs to be close to hand, a mere click away, and everything should be quick and easy. To get the best results from information and services we need to specify our preferences and other personal information. To do this, we have to have confidence in the receiver. Confidence grows over time, so we are more likely to trust a receiver if we have done business many times before. "Closed landscapes" are more attractive for less technically oriented users that want simplicity and trust. The strength of "firewalls" is that they can bring a modicum of order to the chaotic and enormous offering of the Internet.

In Chapter 12 of this book we have developed four scenarios illustrating possible ways in which the mobile marketplace might develop. The scenarios "Professional Users" and "Community Lifestyle" are early-adopter scenarios. Professional Users are likely to be a little more stringent about privacy issues than the Community Lifestyle users. These two groups, probably existing in parallel, represent a fairly limited market. However, with a more positive development of content and technology the market might equally move towards a mass-market phase. "Mobile Klondike" and "Trusted Guide" are mass-market scenarios. The Trusted Guide scenario is one in which closed solutions dominate, but in the Mobile Klondike scenario open-access solutions are the norm. If we look beyond a five-year period it is likely that open-access solutions will be striven for. As new systems become established, users will grow familiar with the new technology and get more and more adept at sorting and browsing through complex information. If the mobile marketplace is ever going to be as successful as many players in the market are hoping, there needs to be integration between virtual and physical markets. This will probably only happen in an open landscape.

Twenty consumer trends

Individualism

Individualism is one of the most important driving forces of postmodern society. Individualism is all about the separation of the individual from the collective; it is not only about individual rights but also to a great extent about responsibilities. As individuals we can be held responsible for decisions that earlier would have been made on our behalf by the collective. Education and pension schemes are two areas, which earlier were handled collectively and applied equally to everyone. Today, we have to make individual choices about what schools to send our children to, and where to invest our pension contributions. Greater significance is attached to the notion of separating ourselves from the herd, cultivating our own interests and styles irrespective of the social contexts we live in. To some extent, perhaps, the social pressure to conform has lessened. The dream of freedom and movement – another important metaphor of the mobile marketplace – is also closely associated with individualism. More and more time and effort is devoted to self-realization. The media is stuffed with messages about self-realization. Research shows that today's young people place as much importance on having an interesting job and pleasant colleagues, as they do on high salaries.[55]

"High tech – high touch"

In his 1986 book *Megatrends* John Naisbitt wrote about the phenomenon known as "high tech – high touch." The meaning of this phrase is this: the more technology in our lives, the greater our need for physical meetings, real experiences and sensory experiences. Technology can never replace the importance of sensory experience. In the 1960s, it was generally believed that within a few decades people would be popping a few pills every day rather than eating proper food. However, the "high tech – high touch" phenomenon directly refutes this type of development. Even though electrical communications are massively on the increase, people also travel more to meet colleagues in the physical world – colleagues that otherwise would be known only in a virtual environment. The desire to find antidotes to our technological day-to-day reality can be discerned in popular pastimes such as interior design, DIY, gardening and New Age spirituality. Naisbitt himself explained "high tech – high touch" as recognition of the fact that "art, storytelling, play, religion, nature and time are equal partners in the

The Individual

evolution of technology because they nourish the soul and fulfill its yearnings."[56] The mobile marketplace can be a fully fledged complement to the physical world, but without a doubt they will both exist in parallel.

I ... (free choice, individualization, multiple identities)

As we have already suggested, one of the most important of all paradigm shifts has been the freeing of the individual from the collective. The process has been under way for hundreds of years, beginning in the Renaissance, strengthening during the Enlightenment and accelerating fast in the 1900s. It is intimately connected with postmodernism.

People are more and more concerned about being seen as individuals, both in the social aspects of their lives and their consumption. To be broadly categorized without any freedom of choice is simply not appealing – most people will turn their nose up at the very idea of it! The expectation nowadays is to be treated as what we are: autonomous individuals with our own preferences, who demand freedom of choice. Shifts of identity can happen at any time and with lightning speed, sometimes several times per day. Hence, focusing on the consumer's own preferences is a big challenge for the media players and senders. The product not only has to be customized to individual needs, but the ways in which communication is maintained with the consumer will vary to an ever-greater extent. The mobile networks of the future, with features such as positioning, will be able to satisfy individual preferences to an ever-greater extent. The possibility of conducting one-2-one marketing backs up this trend.

$7 \times 24 \times 365$ time-efficiency and the 24-hour clock

One of the most serious problems of contemporary life is a shortage of time. An overriding majority of Westerners say that they experience a shortage of time.[57] The hopeless struggle against time means that people are always trying to maximize available time, frequently by means of technical aids. Simultaneously, technical development has turned up the pace of living even more, and any free time that people do have is quickly filled with more possibilities served up courtesy of the possibility revolution. Meanwhile, there are so many ambitions everywhere: to have time to experience things to the full, to be a perfect father or mother, lover, colleague and boss ...

The pace is speeding up, but mentally and physically it can be hard to keep up. The speed of modern life means that more and more people have to take time off work because of mental exhaustion. The annual cost in the EU of work-related stress illness is some US$16 billion.[58] Fast food sales are on the way up in Sweden. With so much to do, who has time to cook? With the increase in IT solutions over the last few years, more and more of us seem to expect to be able to do anything whenever we like. The demand for more flexibility and time-efficiency leads on logically to an expectation that business should be open round-the-clock. Mobile services are an expression of this trend. The Internet never sleeps – "always there, always on." $7 \times 24 \times 365$ is one of the strongest driving forces behind the realization of the mobile marketplace.

The multitasking trend

One way of maximizing time is to do several things at once. We are getting better at this all the time. For instance, we are becoming expert at using several media simultaneously: talking on the phone while surfing the net, working on the move, making a few calls on the mobile while grabbing some lunch on the way to a meeting ... It is not only technical innovation that makes this possible – but also increasing mental bandwidth. There is much evidence to suggest that we are getting better at absorbing large amounts of complex information, because of training from an early age.

At the time of Gutenberg, on the whole people absorbed information that was already known to them (that is, the Bible) and only met a handful of other people in a typical week. With the arrival of the book, people began to absorb new information – that is, information they had no prior knowledge of. Today, we deal with unknown information on a massive scale and come into contact with hundreds of new faces every day. We scan, sort and surf more and more. Elderly people often feel a measure of confusion when faced with the Internet, the sheer volume of available information seems daunting to them. Their generation was based on the idea of *amassing* information, whereas the contemporary focus is on discarding or *sorting* information. The mobile can be an instrument that helps us achieve simultaneous actions. Trends seem to point to a service-based content focused on making day-to-day life easier, while also enabling multitasking.

Relevance

An increasing amount of media and information is competing for the consumer's attention, with the latter plagued by a feeling that there are not enough hours in the day. Thus it becomes more difficult to catch the consumer's attention. The receiver ends up being extremely demanding about only receiving information that is relevant and accessible. The keyword for this can be expressed as FACE: freshness, accessibility, customization and exclusiveness. In other words: the information should be recently updated; preferably it should be exclusive to me; it should also be easily accessed and customized to my unique situation, both from the point of view of the best possible use of my time and how the information is actually delivered. If I am in the car, the news should be read out aloud rather than delivered as a text message ... and so on.

Ultimately this means that consumers become more focused on their own needs, and as a result prefer to choose which advertising they are prepared to receive. Marketing is well on the way to becoming a rather humble form of communication. Today's companies are themselves creating these attitudes by asking consumers to divulge their interests and other personal information. As a consequence, consumers expect to receive relevant information. After all, they have been asked to be specific, and they have been specific. The key to relationship marketing, in this case, is the continuously rising expectations of the consumer.

Reduced decision-making

The possibility revolution increases our available choices and freedom, but choice (as the existentialists have pointed out) usually leads to worry. Quite simply, we are not capable of making an infinite number of decisions.

In other words, freedom of choice has a flipside: we do not have the energy to make the choices. Not only are we having to make increasing numbers of decisions in our decentralized working patterns; even day-to-day decisions take up a lot of time and energy, scarce resources for more and more people. Finally, a ceiling is reached beyond which no further decisions can be made. The consequences of this are already becoming evident – companies are starting to help consumers make those decisions. Agency and advisory services are on the way up, helping consumers find the right electricity supplier, the best pensions policy or maybe just a range of products available out there in the burgeoning market. Getting this kind of help as an individual can make a big difference: avoiding some trivial decisions, having the complexity taken out of others, or being presented with customized or packaged solutions. This takes Big Brother out and puts Mr. Goodguy (your friend on the net) in his place.

Consumers are also deferring purchases more and more while they make their minds up; this means that the marketing has to keep focusing on the customer for longer, to ensure that the transaction actually goes ahead.

Tales and experiences

One of the strongest driving forces for today's consumer is experience. The experience industry is growing strongly and more and more power is focused on the area surrounding the product – that is, on the marketing or the context of the product. Experiences can be disseminated in a variety of ways, for instance by the aesthetic appeal of something, or through the telling of stories. It might just be the case that in a fragmented world people are longing for the unique meaning created by narrative. Stories create a sort of emotional context that adds empathy and reduces complexity. An experience is by its very nature something immediate – it cannot be saved for later. The consumer becomes a part of the product. If the story links to some sort of experience or situation, it will be easier for the consumer to later rediscover that experience. The story becomes a context-marker.

The mobile marketplace is a relatively impoverished substitute for the physical marketplace, and thus it seems likely that the available services will be strongly focused on useful features such as information gathering and communication. The limited data speeds of the networks over the next years will affect the available content, and it will be difficult to provide the user with experience-oriented services.

So what! Surprise me, then!

>Been there, done that, doing it tomorrow.[59]

Most of today's thirty-somethings have in many respects already experienced more than their parents. People are becoming more and more difficult to surprise – the threshold of what actually surprises us or catches our attention is getting higher. Never before have people had access to so much information, or been so widely traveled or thoroughly educated. The impact of the media means that we take part in situations and events all over the world, while thousands of snippets of information daily lay claim to our attention. The paradox is that people nonetheless yearn for renewal. People want to be surprised, and their demands and expectations are constantly on the up. After all, what could be duller than a totally predictable existence?

To gain the attention of the consumer is essential, but to interest him or her is crucial. Commercialization has raised a whole generation of blasé consumers. The consequence of the "been there, done that" trend is something that might also be known as "the-thousand-times-better" phenomenon. What this means is that more and more is expected of new products before consumers are prepared to go out and buy them. The constant upgrades in the market do not offer significant differences in performance and function – it is usually a case of exterior adjustments that are presented as progress and development. The strategy is to make the consumer feel "out" and thus buy new products to feel "in" again. "The new mobile has to be a thousand times better than the one I already have. Unless I'm tired of it …"

Homo Consumentus

I consume, therefore I exist! Consumption has become a way of life for many people, not least in the Western world. More than ever before we are conscious of being consumers, and our self-image is increasingly linked to the idea of consumption. Almost everything we do is based on consumption in some form: of experiences, emotions and lifestyles as well as products and services. In research, consumption is divided into many different areas; for instance, lifestyle consumption is subdivided into areas such as nostalgia consumption, routine consumption and emotional consumption. The clearest new trend is our awareness of our role as consumers. We become more and more clever at interpreting and decoding advertising messages, as well as insisting on our rights as consumers. It cannot be seen as a coincidence that the mobile marketplace is beginning to emerge in the consumption-focused society we live in today. The mobile marketplace is an ever-present arena for consumption, which suits Homo Consumentus perfectly well.

Increasing receptiveness to technology

Openness and technophobia lessen as we use technology more. As we have already seen, it takes less and less time for new technology to establish itself in society. GSM phones, for instance, took only four years to reach 50 million users.

Every generation has its technological references. Today's children are growing up with computers and mobile phones – and using these devices throughout their lives will be unremarkable to them. Habits that seem strange to the parents may end up being ordinary to the children. It is not only IT that affects our receptiveness to technology. The inroads of technology into the public domain add to our increasing technical maturity, and technology becomes an accepted part of our lives. Voice-activated telephone queues, digital and satellite television, intelligent homes, the healthcare system and genetically modified food are all things to which we are daily exposed.

However, new technology is not immediately acceptable to everyone. As we have seen, the introduction of new technology is often taken up in the so-called S-curve pattern – with visionaries and pioneers initially venturing into the market followed by pragmatists, conservative-minded consumers and the mass market. The curve illustrates an interesting dilemma that is well known to most technology-based companies. Geoffrey Moore[60] called it "the chasm." On one side of the chasm are the enthusiasts or early-adopters. Facing them on the other side are the pragmatists, the conservatives and the skeptics. To build a bridge over this chasm is the crucial step in establishing new technology in the mass market. The big unanswered question in relation to the mobile marketplace is whether this bridge will be built at all.

Convenience

More and more products are being adapted with convenience in mind – longer or 24-hour opening hours, mass-customization and convenient solutions are all being made available to customers suffering from a lack of time. In the physical marketplace, the increasing emphasis on convenience is unmistakable. These days, even some shops seem to offer comfortable sofas, coffee machines and newspapers. A recent advertisement for the Volkswagen Passat asks: "Have we made it too comfortable?" And then adds: "The car is much safer and not least, much more comfortable." The advert then goes on to inform people of the new specification, which includes climate control, remote control boot and a refrigerated glove compartment. In other words, the car is being sold on convenience, which sets the car apart from many of its competitors where performance and safety are prioritized.[61] Today, people expect their requirements to be met quickly, with high quality and convenience – without any trouble. Technology is still plagued by all sorts of complications from a user perspective. As a result, research into user-friendliness will be prioritized in the future.

Conditional loyalty, mutual respect

Brand, brand, brand ... This is the clarion call of marketers. Yet the truth is that as consumers we are less and less loyal to brands. We are not loyal to one supplier, but to many suppliers in parallel. This is a direct consequence of the postmodernization of society.

As a consequence, companies can no longer take brand loyalty for granted. The consumer will be loyal insofar as it suits his or her own rules and dictates. Like teachers and leaders, brands will have to earn loyalty to the extent that they deserve it. "I buy from a supplier that can deliver what I want, when I want it, at a price that suits me. I buy from someone whose style I subscribe to ..." And just like a teacher or leader, the loyalty has to be based on mutual respect.

The individual's changeable sense of identity and lifestyles means that he or she identifies with different brands in different situations. Volvo's slogan "For Life" suddenly looks set to get a drubbing from a competing, more immediate slogan such as "For Now."

In spite of this, the value of the brand is increasing in a virtual marketplace where the various players are often separated by large distances. To have confidence in the sender creates a sense of security – and this will be a major factor in long-distance commerce.

Professionalism

What we mean by professionalism is the increasing demand for precision in adapting to individual needs. As consumers we are irritated by any failure to quickly and appropriately satisfy our needs and requirements. Research has shown that e-commerce suffers from a lack of professionalism. Eighteen percent of consumers report failures in delivery of products, and 78 percent indicate that they have been subjected to delays or other irritations in the purchasing process.[62]

For the mobile marketplace to establish itself powerfully the players have to be able to satisfy the needs and requirements of the consumer, to ensure that the technology functions as designed, that deliveries are made on time and that information is correct. Very high expectations are placed on the player or supplier that interacts with the customer. The consumer is very particular, and very conscious of quality.

Privacy

One of the most uncertain but also decisive trends for the mobile marketplace is the whole issue of privacy – personal integrity. There are two dimensions to the question. How much information about ourselves are we prepared to divulge? And second, there is the whole Big Brother argument: What can the information be used for, might it be used against me or in a way that goes against my wishes?

Without a doubt there has been some integration between the personal and public spheres over the last few decades. As individuals we have become willing to allow aspects of our private lives to come under official scrutiny. The media has also been moving steadily towards increasing news coverage of people's private lives. In the autumn of 2000 there were no less than five documentary "soaps" on various television channels in Sweden. When the so-called Monica Lewinsky affair was rolled out as a major international news story, we were able to get a blow-by-blow account of President Clinton's conduct. There was even discussion about whether the President should be obligated to show his penis to the Court. A few years ago it would probably have been inconceivable that this kind of story would ever make front-page news.

We all leave traces whenever we use the Internet, and with the help of positioning it is already possible to identify the geographical position of mobile phone users. If people were more conscious of being watched, they would probably be more careful about their Internet behavior and the sort of information they were prepared to hand out. The unique factor of the mobile marketplace is that position services will be possible; but the question remains: Will the consumer allow positioning?

Increasing advertising exhaustion

One way of shielding oneself from online advertising is to subscribe to services that delete personal information. Other options are to block advertising to the mailbox. In the electronic world it is also becoming more and more common to have more than one e-mail address, so that valuable information can be kept separate from viruses or unimportant information.

The phenomenon of growing advertising exhaustion might simply mean that people are not prepared to receive "push services," and would rather themselves decide when and what sort of advertising messages they are prepared to receive. As experienced consumers we are getting better at seeing the difference between good and bad advertising. In particular, younger generations seem to have an inbuilt "bullshit indicator." The demand for relevant and exciting advertising will intensify even more in the future.

More consumer power through virtual communities

IT technology has made it possible to organize new consumer groups and in this way increase the power of the consumer. This gives the consumer the power to act in his or her capacity as Homo Consumentus Politicus. Today there are many sites up and running such as PlanetFeedback.com, which in various ways increase the influence of the consumer, while also acting as a meeting place for consumers to exchange experiences and knowledge. When information technology was still in its cradle there were many doomsters who believed that it would lessen the power of the individual, whereas in fact the opposite has happened. For instance, with cheap and easily available technology it is a simple matter to set up a homepage. Millions of chat rooms all over the Internet are an instant and fertile breeding ground for all sorts of gossip, both of the true and false variety. Virtual communities spring up like mushrooms all over the place, in many instances financed by the online hosts. These are an excellent way of getting streetwise help with product development and marketing; and listening carefully to these virtual discussion forums can be very profitable for companies. For instance, Bruce Springsteen's manager, Jon Landau, recently reported that he had secretly kept an online watch on the Boss's fans and, as a result of what he learnt, made changes to the repertoire and ticket prices, as well as to future projects – all to live up to the fans' expectations.[63]

More people are buying digital assistants

With the consumption of new hardware such as PDAs (personal digital assistants, handheld computers, and so on), people are making themselves dependent on technology to carry out their daily work, make phone calls, book tickets, organize their calendars, archive documents, and so on. Only

2 percent of Sweden's population own PDAs, but another 6 percent plan to buy them.[64] Gerry Purdy, CEO at the research company Mobile Insights, believes that more and more companies include the PDA or palmtop device as something integral to office equipment; to be able to upload the company's database will be as important as adjusting the calendar. There is ample evidence to suggest that it is primarily corporate users that will acquire digital assistants. But hardware development is leading to a greater integration between computers, telephones and digital assistants. Integration will soon be the norm. The question is, how many new users will flock to PDAs?

From male to female

At the end of the 1900s, big value shifts had set in which affected the relative roles of men and women. There is, for instance, much more diversity both in families and sexuality. The old, obvious roles are starting to look outdated: man as breadwinner, and woman staying at home taking responsibility for the home and children. Women have higher purchasing power than ever before, and play a much more influential role in the workplace. Women are also responsible for most of the family's purchasing decisions.

Futurologists such as Warren Benis, Alvin Toffler and John Naisbitt believe that women are better equipped than men to lead tomorrow's organizations. Leadership will be based on emotional intelligence and the ability to create and maintain relationships and collaborations. Among postmodern consumers there is an expanding group (also known as "Cultural Creatives"), usually comprising more women than men. Six out of 10 in this group are women. In the US they make up 24 percent of the population – in the US and Europe there are somewhere in the region of 50 million Cultural Creatives. A whole range of values distinguishes this group from others. As a rule, Cultural Creatives are less interested in money, and more concerned with self-realization and authentic experiences. They are well-informed and critical consumers, often prepared to pay a premium for environmentally friendly products or services. They are more likely to be "information junkies." They prefer to read and listen to radio; they scan media sources for interesting subjects before deepening their knowledge. The group does not necessarily buy the latest computer equipment or mobile phones. However, it is at the forefront of many cultural innovations: CCs tend to be innovators and opinion leaders for some knowledge-intensive products.[65]

The Individual

Increasing purchasing power among women, and value shifts in favor of environmentalism, feminism, recognition of global problems and spiritual development, obviously mean that products and marketing will increasingly be targeting this group.

But how many of today's IT companies are run by women, and how often are female consumers the main point of reference?

Chaos on the Staircase of Life

This is the name of a project run by the Swedish Institute of Future Studies a few years back, referring to the eclipse of an old model – an old staircase – and its replacement with a new one. The apex of life is no longer at 50. One example of this new staircase is that many (men) are starting afresh at 50 – with new children and new families. Another example is what one might call the teenage syndrome – today, anyone from three to 93 might be wearing the same clothes or listening to the same type of music. There is also talk of a new life phase, known as the "preteens." The consequences of these factors are clear: it is becoming more and more difficult to use age as a basis for segmentation.

SUMMARY

- The net transfers power from the seller to the consumer.
- Postmodern, pluralistic values are forming the modern individual and leading to a desire for choice and personal touch.
- Worldliness, high expectations and the professional consumer are putting more and more pressure on suppliers to perform in areas such as delivery, relevance, convenience, as well as building loyalty among their customers.
- The "possibility revolution" is providing people with more and more choices, thus making it imperative to sort and decline information.

CHAPTER 8

The Economy

What is technically possible and required by humans is not always economically realizable. In this chapter we are going to look at the economic prospects for the mobile marketplace, as far ahead as we can see. The unknowable factors are many, as one might expect – we have already touched upon some of the questions of the economics of the mobile marketplace.

What we will concentrate on here, then, are two main questions. What size of investments are we talking about? Where can one expect the revenues to come from?

Gigantic investments

The development of the mobile marketplace is an industrial project of mega-size. Before a single service has been delivered in the third-generation mobile network, investments in the order of US$300 billion will have to be made over the next three to four years. Critics believe it will take several years before these investments can be recouped. Industry players are more optimistic, and believe that within a few years the market will be worth some US$200 million per year.

Development of systems and telephones

As an industrial project the level of commitment for third-generation telephony is almost in a class of its own. Ericsson alone is investing in the region of US$2.5–3.5 billion between 1999 and 2001. To put this in context, this is US$0.5–1 billion more than the total cost of developing the platform on which Volvo S80, V70 and S60 are based. Another comparison: AstraZeneca's best selling drug Losec cost US$500–600 million to develop.⁶⁶

High development costs are engendering many joint projects and anticipated mergers in some industry sectors to reduce R&D investments. This also explains many of the joint ventures we have seen in recent years.

Investment in networks and licenses

The 3G project is also enormous as an infrastructure project. Yet the buoyant hopes of industry for third-generation telephony made recent license auctions in the UK and Germany exceed all expectations. In the UK, each license cost 4.5 billion GBP and in Germany 16.5 billion D-marks. In other words, each license ended up costing around US$500 per head of the population. In countries like Sweden, where the governments have opted for "beauty contests" rather than auctions, critics have complained that the state has recklessly squandered this opportunity of filling the public purse. However, after the UK and Germany, bidding has been sluggish, and the license auctions held in Switzerland, France, Belgium, Poland and Italy might almost be described as fiascos. In Belgium, the government had hoped to raise almost 1.5 billion euros from the sale of four licenses; but the final sum was less than a third of that, with the fourth license unsold (Table 8.1).

However, there are also pure infrastructure investments to be made in networks. The 3G networks, known as "fine nets," need a high density of base stations in order to work. In areas of sparse population such as Scandinavia, considerable investments will hence be needed to achieve coverage outside of the main urban areas. One possible way of handling this problem is for the operators to join forces, as in fact they have already begun to do. Through collaboration between the three winners of the Swedish "beauty contest," the investment prognosis for the expansion of the Swedish networks has been slimmed down from 100 billion SEK (US$10 billion) to 30 billion SEK.

However, this will probably not be enough to get sufficient profitability on the investments. In spite of substantial investment savings packages, analysts at the consultancy Northstream calculate that a profitable investment would necessitate subscriber spending of around US$90 per month. Current average monthly spending is closer to US$40.⁶⁷ Even for well-established operators this looks like a tough proposition.

In 2001, economic factors have become a major obstacle to the development of 3G. The reasons for this are clear. Operators that have participated in the European license auctions have run up large debts and still have major

Table 8.1 License methods, the number and costs

Country	Method	No. of 3G licences	Average prices per license
Australia	Hybrid	6	US$5–148 million
Austria	Hybrid	6	100 million euros
Belgium	Auction	4 (3)	150 million euros
Denmark	Auction	4	US$118 million
Finland	Beauty contest	4	Free
France	Hybrid	4	2.6 billion euros (June 2002)
Germany	Auction	6	8.4 billion euros
Ireland	Beauty contest	4 + 2	No decision as yet
Italy	Hybrid	4	3 billion euros
Japan	Beauty contest	3	Free
Poland	Auction was cancelled, existing GSM operators were given UMTS license	3	US$613 million
Portugal	Beauty contest	4	100 million euros
Singapore	Hybrid	3	50 million Singapore dollars (US$28.5 million)
South Korea	Beauty contest	3	US$963 million
Spain	Beauty contest	4	125 million euros
Sweden	Beauty contest	4	US$112 million
Switzerland	Auction	4	34 million euros
The Netherlands	Auction	5	550 million euros
UK	Auction	5	4.5 billion GBP (7.2 billion euros)
USA	Regional auctions		–

Source: Various

commitments. The European operators have spent and are spending about 300 billion euros. They are already indebted to the tune of some 240 billion euros.

Operators such as Telefonica have been highly proactive in 3G bidding with their Finnish partner Sonera. Other players such as Vodafone, British Telecom and Deutsche Telekom – which made bids for 13 licenses in the European license auctions – have all similarly built up enormous debts. Worst off are France Telecom, Deutche Telekom and British Telecom with debts in the region of 40–60 billion euros.

The operators had planned to finance these investments with new emissions, but with the collapse of the world share market during 2000 and 2001, this became impossible for most. Joint ventures in network expansion, loans from EIB (the European Investment Bank), supplier loans from system providers such as Nokia and Ericsson, or even repayment of license fees are possible routes for solving the operators' financial crises.

Investments in network expansion do not only affect mobile operators and systems suppliers. The construction industry is also deeply involved in the expansion of both fixed-line and mobile broadband. About half of the investments in new networks are expected to end up in the pockets of the construction industry. The construction company Skanska, which is Ericsson's project partner, may earn up to US$32 billion in a period of five years, if Ericsson pick up 40 percent of the orders on the 3G network.[68] The profit margins for this type of construction project are also higher than for some other projects.

Where are the revenues?

In other words, today's big question for the mobile marketplace is about revenues: where is there money to be made for mobile operators, service providers and content providers? From call charges, payments for products and services, sponsorship? From transaction charges levied on m-commerce (mobile shopping)? Or, from advertising? Or ... ?

The bottom line, of course, is this: will there be enough revenue? Arthur Andersen, the consultants, stated in a report in October 2000 that it would take 15 years for the existing operators to recoup their investments in 3G networks.[69] Other analysts make similar predictions.

Judging by the experience of Japan's I-mode, it is possible for mobile operators to develop combinations of revenue streams. This is going to be essential, as call charges are expected to come down. In the case of I-mode as we have earlier seen, the revenues consist of call charges and a further monthly fee of US$2.5 for unlimited use of certain services. Extension packs cost another US$7.5 per user per month. Added to this, DoCoMo makes a transaction charge on all companies selling services in the system. And finally, call volumes increase because the average I-mode user makes more calls than a "normal" user. A similar model was, in November 2001, chosen by Swedish Telia in a settlement with content providers.

M-commerce is another potential income source. Success in the mobile environment will call for a great deal of fresh thinking. Scaling down the Internet sites to a format that fits the mobile display is not enough on its own. An analysis carried out by Accenture[70] has shown that it takes two and a half times more button pressing to buy a CD from Amazon in a mobile environment than via fixed-line access.

There is a great deal of mixed opinion on the potential of advertising to generate revenue streams. In a 2000 Forrester brief,[71] the advertisers predicted that 18 percent of the online advertising market would be mobile by the year 2005. Forrester's own analysts were less sanguine, and believed the market would amount to no more than a fifth of this. The rationale for this lower estimate was that most people regard the mobile phone as a communications and productivity tool. If advertising gets in the way of these services it will not be accepted. Furthermore, the amount of time spent online will be less than with fixed-Internet use, hence limiting exposure.

Research also showed that 50 percent of PDA users were not at all interested in receiving mobile adverts.

Forrester's findings were well in line with the investigation into Swedish users carried out by Kairos Future in autumn 2000. Communication and productivity services are entirely dominant, along with entertainment services among younger users.

However, positioning will create opportunities for customized, position-specific advertising. This might be a route for an entirely new kind of "push-advertising," which consumers can either accept or refuse as a form of consumer information. Even if only five percent declare an interest today in this form of advertising, this may not be significant in terms of making predictions in an up and running mobile marketplace.

Two markets and two consumer groups

Many analysts agree on the fact that the market for mobile services will divide into two principal markets. The first will be business-to-employee – used by companies to expand their local networks, keep in touch with sales personnel and consultants, and so on, using mobile phones and intranets. The other market will be the consumer market, as shown in Figure 8.1.

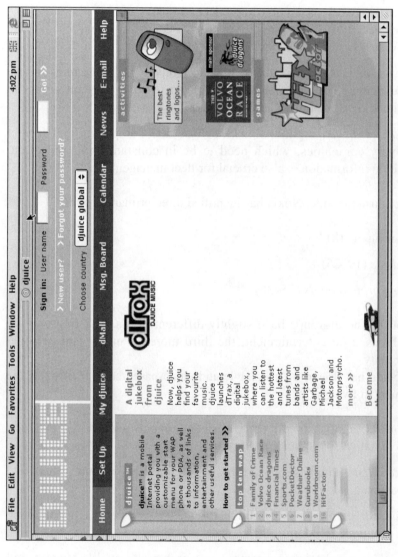

Figure 8.1 Norwegian Telefonor's mobile service DJUICE targets youth in Scandinavia, New Zealand, Thailand and Singapore

Source: www.DJUICE.com

Nokia sees three primary customer groups in the corporate segment[72] and others make the same assessment (Figure 8.2). These are:

- *Sales-driven* organizations, such as manufacturers and banks, where sales personnel in the field need to be permanently connected to the internal networks, to be able to produce estimates and calculations on the spot, check orders and provide support for consultants, and so on. Here, CRM systems will be central.
- *Service-driven* organizations such as consulting companies where "knowledge employees" in the field need to have constant access to methodology, resource planning systems, and so on.
- *Logistics-driven* organizations, such as taxi companies and freight delivery companies, which need to be in constant communication. Position information is also crucial for fleet management.

On the consumer side, Nokia has identified three primary groups:

- *Teens* (up to 18)
- *Students* (19–25)
- *Young professionals* (25–36)

Obviously, these groups have slightly different needs. The first two are more focused on entertainment, the third more on utility and personal productivity.

Ericsson has drawn up a lifestyle and value-based segmentation of the customers in five groups:[73]

- *Pioneers:* emphasis on empowerment, experimentation, sexuality, individuality, spontaneity and action.
- *Materialists:* emphasis on creativity, vitality, body culture, status and appearance.
- *Achievers:* emphasis on flexibility, rationality, success, positive thinking and economic freedom.
- *Sociable:* emphasis on harmony, environment, quality, health and personal privacy.
- *Traditionalists:* emphasis on order, stability, hierarchy, security, control and simplicity.

The Economy

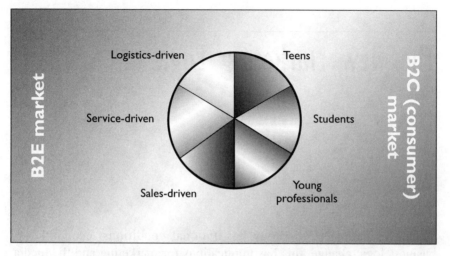

Figure 8.2 The market for mobile services
The future market is expected to consist of two dominant submarkets: business to employee (B2E) and the consumer market (B2C), with different core groups. The size of these, however, is not clear
Source: Nokia

SUMMARY

- Gigantic investments, uncertainty about business models and revenue streams.

- Data and voice communication, sponsorship and advertising, sales and delivery of utility and entertainment services are the activities that will generate revenues to finance the networks. The relative importance of these is unclear.

- Mobile intranets are a natural market with great potential for sales-, service- or logistics-driven organizations.

- Young people and young professionals are the two main consumer groups.

CHAPTER 9

The New Marketing Logic

Whenever people, technology, institutional conditions, economics or business logic change, this has implications for marketing and the media. Marketing and the media play no less of a central role in the mobile marketplace. From the perspective of media and marketing, the mobile phone is a new channel and is already starting to be used as such. How it is used in the future will largely depend on its function in the changeable media and marketing landscape.

In this chapter we begin by summarizing the three major trends: the development of a customer-driven marketplace, the emergence of micro-geographic marketing, and position commerce.

Next, we make in-depth forays into changes in media and marketing.

The customer-driven marketplace

Over the last decades we have slowly moved away from product-driven markets where products and production have been the driving forces, to distribution-driven markets where marketing and channels had great influence over the information process in creating, controlling and retaining brands and brand loyalty. Both of these phases were characterized by a weak loop-back to the marketer. The marketer had a limited, static understanding of the consumer's needs and preferences. Distributors and media channels had great power over the flows.

The new marketplace that we are heading towards is known as the *customer-driven market*. The Internet brought another dimension to the marketing logic – interactivity. The new marketplace brings even more opportunities of connecting the loops; as well as the possibility of quantifying customer needs and preferences 24 hours per day. This knowledge, gathered by marketers, media players and distributors has a controlling effect – dynamic rather than static – the importance of which is growing dramatically. The complexity of proliferating media channels will effectively mean that the one-sided power of information – that is, the power of the broadcaster – is in the process of being smashed to pieces. To create a viable competitive edge in this new system, differentiated value regimes have to be created in which the consumer is preeminent. If loyalty and brand positions are to have any longevity, it has to be realized that marketers, media channels and the consumer must create value equally. This might also be known as shared responsibility.

Interactivity can be taken a step further in the mobile marketplace. The reason for this, as we have seen earlier, is that yet another dimension can be added in the form of customization and time specificity – that is, the possibility of acquiring information at the exact point in time when it is needed. This third dimension could be termed "spatial specificity."

Let us briefly examine the practical implications of this change.

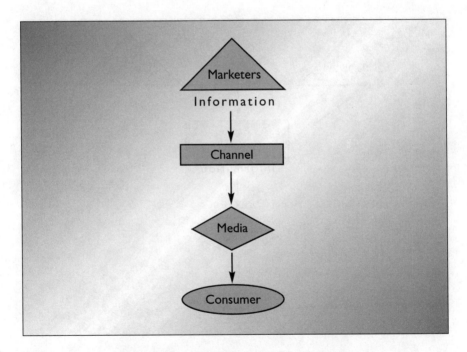

Figure 9.1 The product-driven system
Source: Reprinted from the *Journal of Advertising Research.*
Copyright 2000, by the Advertising Research Foundation

In the product-driven system, companies created products or services that generally could not be easily copied or subjected to excessive competition. The distributors were generally few and dominant in their categories. This meant that, in practice, successful manufacturing companies dominated the market – including channels, medias and the end-customer (Figure 9.1).

In time, the market evolved into a distribution-driven system. It became more and more important to offer a wide choice and range of products at designated points where the customer could make purchasing decisions. Distributors took power into their own hands and controlled the marketing organization and information channels. Proximity to the customer and information about the customer also made it a natural progression to exert control over the information. Media channels were largely affected by the same thinking (Figure 9.2).

However, the customer-driven or interactive system is characterized by the integration of all involved parties. Power is pushed towards the end-customer. This has happened predominantly because of changes in the

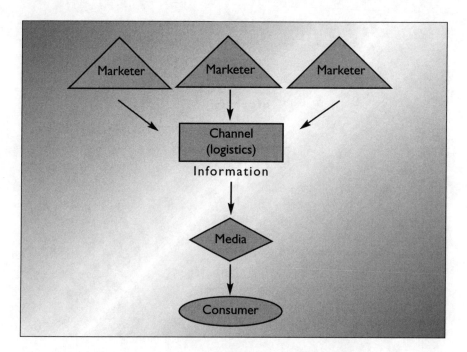

Figure 9.2 The distribution-driven system
Source: Reprinted from the *Journal of Advertising Research*. Copyright 2000, by the Advertising Research Foundation

The New Marketing Logic

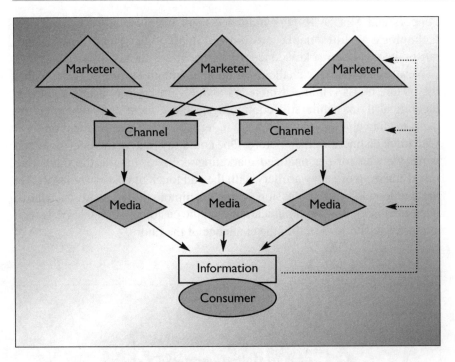

Figure 9.3 The customer-driven system
Source: Reprinted from the *Journal of Advertising Research.*
Copyright 2000, by the Advertising Research Foundation

information structure – technical developments have placed one end of the communication channels in the hands of the customer. The plethora of various types of media now emerging, have the power to browse and sort and function as transformers and generators between the purchasing power of the consumer and the marketing pressure of the distributors (Figure 9.3).

Broadly speaking, the great marketing challenge (as suggested by Peter Elley, Grey Academy, Grey Global Group Nordic) is based on the total shift in marketing approaches made possible by new technology. At the heart of the new world picture the customer has taken the place of the product. We have to become better at understanding and focusing our marketing efforts on the customer's personal "universe." In spite of this, the general direction of media in recent years has been to merely divide the marketing environment into "online" and "offline" denominations – both of which attempt to grab the customer without any attempt at coordination.

The digital world has created new ambitions and potential for a new era of relationship marketing. Relationship marketing groups such as Don

Peppers and Martha Rogers indicated right from the start that the new technology would enable customized marketing. "One-to-one" and customization became key concepts for the whole marketing field.

New research is now challenging these assumptions, and suggesting that the picture needs refining. Marketers and advertisers repeat this mantra in theoretical terms while still continuing to place most of their money on mass-media vehicles. This is mostly in terms of push-communications, but with more distinctions than before. One could say that, in the interim period while we wait for personalized marketing skills to mature, the marketing mix is made up of heavy artillery with the odd touch of customized skill. Yet the whole marketing profession seems unanimous in its desire to achieve integrated marketing communications and the customer-driven logic is seen as a prerequisite for meeting the challenge of the future.

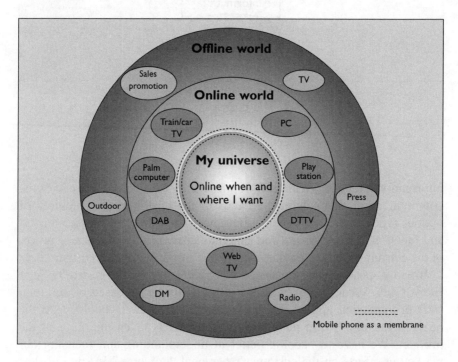

Figure 9.4 The mobile as a membrane between the personal universe and the external world

Mobile device functions as an interactive interface – the customer's personal "sword" and "armor" that protects him/her and builds a protective membrane between the personal universe and the external world outside it. The figure is inspired by Grey Global Group Nordic's platform for communication and developed from analysis (*Deconstructing Media* 2000) made by Forrester research group

After all, the customer is not concerned about how many departments, professions or sections marketing professionals utilize to create reactions among consumers. The customer is acquiring increasing numbers of tools for accessing information whenever it is required, and also to stop the flow of information. Whenever the customer wishes, price comparisons can be made or decisions made on the basis of any extant brand loyalty. The development of a mobile marketplace makes this process even more sophisticated and additionally strengthens the customer's power. There are very compelling reasons why advertising channels and marketing professionals must take this into account if they wish to achieve success at reasonable cost in their communications with consumers. A crucial question increasingly being asked in marketing terms, is how to achieve consumer loyalty in the interactive world?

Customers will in all probability be using the mobile as a membrane between themselves and the information flow. By means of this membrane they can let the world know their "mental state" – that is, what types of information and communications they are prepared to receive in any given moment (see Figure 9.4).

The evolution towards micro-geographical marketing

> There is an old adage in advertising that the closer to a purchasing opportunity you get, the better your opportunity of making a difference as an advertiser. Any possibilities of using media channels that take my offer closer to the customer, are therefore interesting because of the effect my marketing may finally have.
> (Carl Wåreus, Director of Marketing, McDonald's Northern Europe)

Many people believe that the evolution towards micro-geographical marketing can potentially meet the challenge inherent in the fact that consumers defer their purchasing decisions the closer they get to the actual purchase. Seventy-five percent of all purchasing decisions are made in the shop.[74] Excessive numbers of products to choose between and similarities in terms of function and price, make it imperative for the marketer to maximize every opportunity of communicating with the customers during the entire purchasing process. To achieve this, new channels need to be used – channels that are not dependent on time and space, channels that accompany the individual customer wherever he or she goes. Attempts are now being made to send advertising messages to customers when they are facing "the moment of truth" in the geographical marketplace – in the shop or shopping center. Previously, the influence of the distributors was limited

Figure 9.5 Influencing purchasing decisions in the shop

to sales promotion at the point-of-sale. The rest was in the hands of specialists. However, in future there will be a great possibility for distributors to maintain ongoing conversations with customers, right up until the moment of truth in the shop – and thus refine purchasing decision or strengthen customer relationships (Figure 9.5).

Achieving success with micro-geographical marketing is easier in theory than in practice. Any advertiser can see the benefits of customized information or advertising adapted to the geographical situation of the consumer. But an equal number of advertisers, in our study, expressed doubts about how to handle and constantly adapt enormous information loads cost-effectively so that mobile advertising messages achieved the desired effect. Furthermore, they were aware of the fact that most companies are still not utilizing their consumer data effectively, even before two further dimensions – time and place – are introduced.

> Every advertising message costs money. If I know that I am reaching 25 million people by using television, I can devote a certain amount of time and money to make sure that the presentation is good. However, if we are dealing with offers destined for a single store where I might reach no more than 50 highly qualified

The New Marketing Logic

sales leads, I'm not sure how much time I can put into that message or how many sales will be generated as a result. If it's going to be profitable, there has to be some sort of automation that makes it easier for the marketer to work with high volumes in different locations, with similar but adapted offers. (Carl Wåreus, Director of Marketing, McDonald's Northern Europe)

Position commerce

The emergence of micro-geographical marketing is therefore dependent on the eventual development of position commerce. Corporate players in this market maintain that the automation Carl Wåreus is calling for already exists, and the real problem today is that not enough marketers are aware of the possibilities of automated information processing. While technology must be able to distinguish where people are and perhaps also what their needs are at any given time, there will also be possibilities for retailers and geographical marketplaces to buy exact information about customer flows, thus enhancing the precision of their marketing and advertising. This development could in fact restore the physical marketplace's position as one of the most important marketing and sales channels in history.

While this type of service may benefit the retailer, it may also ultimately be disadvantageous, as the consumer (even while in the actual shop) will be able to access price comparisons both in virtual and high street environments. The risk for the retailer is that the consumer will actually make a purchase using the mobile, or select another high street retailer. There are more and more examples of services delivered to operators, mobile portals and directly to the end-customer. Often these services are of a type that might be described as a "sales finder" – that is, real-time location of shops or points-of-sale offering the most competitive prices. The company Mobile Position has developed a number of successful B2B- and B2C-focused applications. Among its consumer products are applications such as Yachtposition, Bikeposition and Friendposition. One of the company's most innovative products is mPass, which gives mobile operators the opportunity of protecting the customer's personal integrity while also integrating selected external applications.

With the help of companies like Mobile Position, Scandinavian operators such as Telia are now establishing position services. In the case of Telia, the position services are what might be described as "near you" services – anything from map services to locating houses for sale, prices of individual properties, and so on. One of the more spectacular services is the new computer game Botfighters, played both in physical and virtual space, which gives it an entirely new dimension. Players create, name and equip their own robots with attributes online. You download the game to your mobile phone. After that, it is simply a question of being prepared, as anyone in the local physical environment might be a player equipped with his or her own robot. The object of the game is either to get within range of one's opponent and strike, or move out of range of the opponent's weapons. The big question is, who in the crowd, in the shopping center or the street, is playing the same game?

(For further information on position commerce, see Chapter 10.)

Media trends

Major changes are currently under way in the media industry, mostly connected with the digitization of information, the emergence of new channels (Internet and mobile Internet), decreasing production costs and new media habits.

We are now going to take a brief look at some of the changes most significant for the mobile marketplace.

Increasing integration of various media

Rapid technical evolution, together with convergence between telecommunications, IT and the media sector, are the foundation for a likely development towards media that select the best of all worlds. The personal computer, the electronic book, outdoor screens with moving images, the web page, the telephone, and moving advertising images in the taxi and on the fridge – all have their own possibilities that can be integrated.

Media convergence and broadband

Distribution costs are in free fall. Transfer costs are approaching zero. More and more media channels are converging, no longer just television programs but also websites, magazines, electronic newsletters, and so on. A good example of this is the British nightclub Ministry of Sound, simultaneously a website, a print magazine and a music label (with its own CDs). Even the traditionalist BBC is making sizeable investments in the online arena and now fields its own online editorial department. Other traditional media such as music and film are also reaching people via the Internet. Film producers are looking forward to cheaper film distribution via the Internet and the much-discussed Napster (napster.com) announced a while ago that it was on the verge of signing an agreement with three giants in the music industry: Warner, BMG and EMI. All three had earlier sued Napster, but later on, interestingly enough, seem to have changed their strategy. Eventually, the giant Sony Corporation announced that it would launch a new Internet service for downloading its musical output. Many other examples of the unbounded digital world have emerged in the wake of the much-publicized Napster court case, which finally led to a shutdown of their business. Millions of file-sharing fans are migrating to alternatives like BearShare and Napigator. The Beverly Hills-based company BigChampagne, which according to *Wired* magazine monitors hard drives and networks, shows that "these underground forums have evolved into mainstream swap meets. As with hit-driven radio stations, the top P2P requests follow new MTV videos and, when broken down geographically, artists' tour dates."[75] OpenNap is a new service, an exact copy of Napster except in one important respect. Rather than placing the company's servers at the disposal of the customer, the customer sets up his or her own server. The entrepreneur Matt Goyer apparently has plans to put a server on an abandoned oilrig – Sealand – anchored off the British coast. The platform was previously the home of an eccentric former army officer who called himself "Prince Roy of Sealand." In the technical sense, the platform is an independent state. At the moment, the company HavenCo is keeping its servers here – therefore it is immune to national or international regulations.

> Peer-to-peer is the next great thing for the Internet. (Lawrence Lessig, *Code and Other Laws of Cyberspace*)

The foundations for media convergence are digitization and dramatic changes in distribution over the Internet – these factors change the media map from the foundations and up. The next big shift, experts believe, will be based on the development of Internet 3.0. This uses new techniques for information distribution, and supports "decentralized collaboration" – defined as a peer-to-peer system for real-time communications. A peer-to-peer system such as the one on which Napster (see above) is based, allows PCs to exchange data over the Internet with little or no mediation by any central computer. In the past two years enthusiasts have used the technology for anything from web searching to parallel processing. Also, in journalism some experiments have been carried out in this area, commonly referred to as Open Source journalism.

All media are digital before they are printed on paper or burned onto CDs (where the content is also digital, of course). Binary codes combine in any chosen way, without difficulty. They can be used and reused together or separately. This, on its own, fundamentally changes the media landscape. As yet we have only seen the beginning of the hybrids now beginning to develop in the borderlands between various media. We have had little time to consider the full impact on media conglomerates that currently dominate the landscape. And we can only speculate about how marketing companies will figure in the restructuring process. New players will appear and this will affect profit margins and alter power structures. However, what we can predict is that media awareness will become more and more important for marketing companies.

Power is being transferred from distributors to consumers

The substantially lower costs of electronic publishing mean that, at least in theory, anyone with a computer and modem is capable of starting a news site or online magazine. There is clear potential for reaching large numbers of people without incurring enormous costs in the process. As a result, 20-year-old computer nerds are not only doing the rounds of traditional media companies – they are also being taken seriously! What they are offering, in fact, are niche media channels.

More and more media are modeled on television

> Technical progress is making online advertising look like TV-commercials, with good-quality sound, images and video in real time.[76]

More and more media are being developed as forms of television content. This is particularly clear in the Internet sphere. Online advertising is quickly moving towards a multimedia style, and according to Jupiter Communications, this is popular with Internet users. In Jupiter's survey "Online Advertising Mix, Scandinavia," more than 70 percent of those polled agreed that online advertising was becoming more like TV advertising. With increasing numbers of people using the Internet and smaller numbers turning to search engines and portals, there is talk among the portals of introducing a channel structure similar to the ones existing on television and in radio.

In outdoor digital poster campaigns, television advertising tends to function more or less as a reinforcement of concurrent television campaigns. This tendency may not be a new phenomenon, but it is becoming even clearer with the latest moving images on outdoor screens. In some public places unique television interfaces have been created, with news flashes adapted to the increased mobility of people in outdoor locations.

All companies want to own their own media channels

Expertise in areas such as customers and market mechanisms is increasing in traditional companies. In view of this trend, BMW is considering whether it really should be handing over 250 million SEK to Eurosport per annum. Rather than Eurosport buying the rights to events that will hopefully attract people synonymous with the kinds of customers BMW wants to reach, BMW will be looking to buy rights on its own behalf, and maybe even broadcasting directly through a leased media company. (Pontus Forström, Editor, *Vision*)

As competition intensifies and products and services proliferate while they also become more similar, companies are increasingly looking to differentiate themselves. They aim to achieve this by developing strategic awareness of the customer. The customer *becomes* the business idea. In order to get as close as possible to the customer, it is essential to monitor customer relations and how they are achieved. As a part of this process, companies are exploring new methods and looking for new media channels. Digital developments in the Internet and internal intranet applications have created a new climate in which companies can take over the function of traditional media channels. The mobile marketplace will strengthen this development.

Traditional media have to find new models

We are moving towards shorter, better and more individualized communications. As soon as interactive television is established (Web TV, digital TV, etc.) we will simply skip any advertising that strikes us as bad advertising. (Pontus Forström, Editor, *Vision*)

In the digital world the old advertising model – with its avowed aim of capturing the consumer in 15–30-second bursts during intermissions in the programming – will be questioned. Experiments are being conducted all over the world to find new ways of avoiding this nuisance.

Increasing demands for speed and precision in the media

Increasing cost-consciousness and drive for profitability have led to more following up of advertising campaigns, and it is now quite normal to monitor campaigns before, during and after their implementation. Increasing numbers of advertising analysts point to the importance of clarifying the relationship between the advertising idea and the campaign strategy. The general public is harder than ever to please, and as a result the advertising is subtler. There is pressure on interactive media to demonstrate tangible results. Hence, traditional media such as print newspapers and outdoor advertising are being scrutinized in the light of competition from mobile and interactive media. But in order to improve precision, monitoring techniques must be quicker and supported by relevant statistics. This is also something to which companies have become accustomed as a result of online media. The general print media lags far behind in this respect. While electronic media offer follow-up within an hour or 24 hours, newspapers tend to report on results about twice per year.

Media consumption is becoming more individualized

One of the clear media consumption trends of the last 10 years is that an increasing amount of media is consumed individually. The days of the family gathering in front of the radio or television are long gone. In the 1950s the popular BBC radio program *The Archers* represented a high water mark of communal consumption – and programs such as the BBC's *The Goon Show* attracted fanatical audiences that would defer toilet visits and leave phones ringing rather than miss a second. Modern media consumers are different in every respect. IT development, globalization and a formidable range of alternatives have changed everything, and the old model of consensus is now completely superseded. Rather than all sitting together in front of the television on a Saturday night, family members are far more likely to be scattered all over the house, each with their own television.

Media consumption is not growing as a measure of time – but we are consuming more media simultaneously.

The New Marketing Logic

The time and money citizens devote to media consumption increase very slowly. An estimated four percent of private consumption is spent on media. In the last 20 years the amount of time Swedes devote to the media has increased by 39 minutes, from 321 to 360 minutes per day.[77] New media tends to redistribute rather than add to the total consumption. To return to the discussion about the introduction of new technology, it is primarily a question of substitution as opposed to innovation. However, our findings do not take into account the fact that several media can be consumed at the same time. In other words, one can simultaneously read a book, listen to a CD and have the television on in the background.

The daily newspapers are starting to lose readers to the Internet, which now occupies the same territory as the traditional evening press. Even television habits are starting to tilt in favor of the Internet. However, the morning paper is still a prominent part of our media habits, and we allocate three times more time to it than any other medium.

One of the main challenges is that consumers are beginning to subdivide into "A" and "B" teams in terms of media consumption. The "A-team" consumer is selective in his or her consumption and prepared to pay for information, while the "B-team" consumer is less dependent on morning papers and tends to watch entertainment paid for by advertising.

The mobile marketplace is changing "the media clock"

There is a clear idea within the media industry about how media is consumed over 24 hours of each day. Obviously, no Mr. Average follows this pattern absolutely, but Figure 9.6 nonetheless gives some idea of the standard 24-hour period.

By studying the diagram we get a clear idea of how the 24-hour cycle has changed. If we turn back the clock a few years, the Internet would not figure in the diagram. Breakfast TV gave consumers in many countries an entirely new medium in the mornings. Portable hi-fi such as Sony Walkmans make it possible for people to listen to music and radio even if they are not driving to work.

To this picture we could add "quasi-medias" competing with the media, such as CDs, video, DVD, PC games, games consoles and MP3 music. In addition to this, mobile manufacturers such as Motorola have developed first-generation so-called "wearables." The idea of this is to integrate the latest mobile technology in clothes, jewelry, bags, and so on.

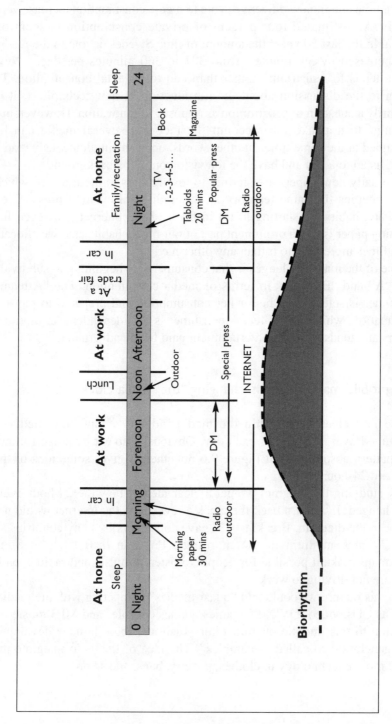

Figure 9.6 The media day
Source: MediaCom, 2000

The question we are interested in here is how the mobile media clock will look a few years from now. And which media will integrate with the mobile – morning papers, radio, traffic news, television, direct marketing (DM), Internet, consumer magazines, books?

The consumer places higher demands on information quality

Consumers are getting used to being served fresh and relevant information – updated by the minute. Both on the Internet and the mobile it is possible to update information as it develops. In this respect, media bound by time factors are falling behind. A typical example of this is classifieds in the dailies already marked as "sold out" – the information is already obsolete before the print run because consumers have accessed the offer on the Internet the evening before.

The factors governing whether or not consumers perceive the information as relevant can be summarized as follows and illustrated in Figure 9.7:

- *Freshness:* "How quickly can I get the information?" Nothing is duller than a morning paper in the evening.

- *Accessibility:* "Is it digestible?" Succinct, clear and pedagogically presented information is more valuable than a nugget of gold in a sea of sand. Amusing and useful information is often more entertaining and valuable than the merely "high-brow."

- *Customization:* "To what extent is the information customized according to my needs?"

- *Exclusivity:* "How many others are getting this information?" This is particularly important if the intention is to make money out of the information. Financial information can be especially valuable, hence legislation against insider dealing.

Freshness and exclusivity are aspects that, potentially, imply that the user can make money out of the information. Accessibility and customization save time.

We are talking about practical, useful information and knowledge here. Clearly, what adds value is *quality*. These variables are not equally important at any one time. If I were looking for a recipe, freshness and exclusivity would be less important than accessibility. If I were looking for financial

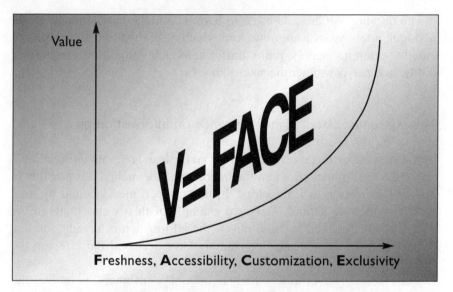

Figure 9.7 Value formula for future media

The value of the information is a function of quality: **f**reshness, **a**ccessibility, **c**ustomization and **e**xclusivity, and the needs of the individual are the determinants of that quality

information, freshness and credibility would be crucial. But if I were a highly qualified analyst, I would be prepared to tackle complex information.

Greater availability of free media

At the same time as information production costs are changing and the media map is fragmenting, alternative financial models are starting to appear. The "almost free" Internet and the traditional advertising-financed models are being combined in endless constellations in the media arena. This development should continue once the mobile marketplace gets air under its wings. Once this happens, consumers will have learnt to attach value to various dimensions of the information flow. Either the information is so exclusive, in-depth or specialized that they are prepared to pay for it; or (if the competition is providing low news quality with piecemeal offerings) they will make tough demands on the balance between advertising and valuable information and on this basis make their choice for a particular portal or newspaper.

Figure 9.9 illustrates one of the central shifts in logic that we predict will strengthen the development of the mobile marketplace. We are assuming that

Figure 9.8 *Metro*, distributed free in public transport hubs all over the world

Source: Metro International

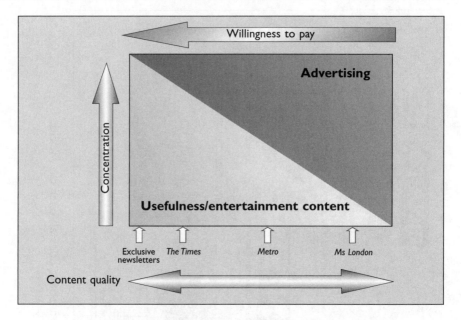

Figure 9.9 Polarization between attractive concepts and exclusive content

The willingness to pay for information has a strong correlation with factors such as packaging, content quality, concentration and advertising content

a polarization will become evident in the mobile marketplace. Packaging (that is, the presentation and format of the information) with an appropriate level of finesse and a complementary range of services will be increasingly important for content suppliers that depend on advertising. At the other end of the scale, it is clear that relevant and focused information will be expected from information services that charge money for their content.

The transition from selling "attention" to selling "accessibility" and "relationships"

Media convergence and changing values and lifestyles may eventually mean that the media companies of the future consist of loose networks of specialists and other participants. Rather than gathering around a channel, they may choose to focus on a specific group of readers attracted to a particular style of presentation. This makes it even clearer that what the media organization (or network) is really selling to the company that wishes to market itself, is accessibility and relationships with a clearly delineated group of readers or users. In practice, this is exactly what virtual communities are already doing on the Internet.

Digital television replaces analogue television

Once analogue television has been replaced by interactive digital television, the media will also have fundamentally changed as an advertising channel. The old-style commercial break will be substituted for a more accurate and informative television advertising strategy focusing on interest groups. Digital television offers niche programing – and therefore the advertising shadows the program content.

The spread of electronic magazines

> If your early morning flight is delayed, that fact should appear as the headline in your personalized newspaper. (Nicholas Negroponte, MIT lab)[78]

More or less all self-respecting newspapers publish their own online edition, and many also deliver news via mobile phones. These digital publications can, to some extent, be customized so that one only receives information one actually wants to read. However, there is a certain nostalgia for the traditional newspaper. At the risk of sounding flippant, the real boost for digital publications will be digital paper making that reassuring rustling sound every time we turn the page!

Marketing trends

As we have seen, marketing is in a process of change. Value creation is being tied to technical evolution and marketing, and hence growing importance is being attached to marketing in the whole equation of success for companies. During the 1990s we saw how brands, CRM, and integrated advertising strategies gained ground.

The growing importance of brands

Advertising and marketing communications have historically been seen as something distinct from the core activity of companies. But in the current climate, where brand values and psychological values are taking on more importance, the core values themselves are being called into question in many sectors. What is the crucial success factor for Ford, car manufacturing or its advertising profile in selected media? The answer is not as obvious as it might once have been. It goes without saying that the product must be excellent and the services ingeniously adapted to the needs of demanding customers. Nonetheless there is a sense that very few cars will be sold unless enough people associate their own lifestyles with the values represented by the brand.

A cursory glance at the largest companies in the stock markets will quickly reveal the high proportion of their capitalized value that is made up of immaterial value. Coca-Cola's capitalization after the debacle in Belgium was in the region of $164 billion, of which estimated immaterial values accounted for $158 billion. Ninety-six percent of the value of the company resided in the brand and the recipe.

The attitude of the brand is ever more important

Modern people have an amazing need – and ability – to clarify their perceptions of self, both at home and work, through association with various brands. These citizens of our post-industrial world have developed the ability of playing with identities and lifestyles. Supporting this process is a whole range of products and services loaded with values and identity spheres that give focus to the kinds of statements consumers are anxious to make about themselves.

For this reason, it is more and more vital that brands have "attitude." The brands have to be almost anthropomorphic – they take a stand, they have values and character and we enter into relationships with them.

Rising costs of gaining new customers

Acquiring new customers is a costly proposition. The information flow is rising exponentially. Brands and information compete for our brains. Traditional advertising communication is based on the idea of interrupting human activity for long enough to be able to catch people's attention and thereby get a message across. But excessive amounts of advertising make us increasingly resilient to these messages. Tomorrow's advertising and marketing must become considerably more sophisticated to overcome this hurdle. Consumers need to feel empathy, and the advertising has to be subtler, more relevant and logical, preferably clever – and interact with our dreams. Advertising has to chime with well-informed consumers who do not want to be bunched in with the rest of the crowd. The mobile marketplace will potentially improve access to the consumer – marketers can follow customers at closer proximity, further down the line towards the purchase. Interactivity and the possibility of holding a lively dialogue with the consumer, result in advertisers having to put more energy into preparation in areas such as target group analysis, creative factors and the budget.

"Storytelling" is becoming a prominent element of marketing

The need to make connections and explain complexity are ushering in a renaissance in the ancient art of generating knowledge through storytelling. Stories are mnemonic devices, creating emotional responses to situations, events and, ultimately, brands.

Marketing communication is moving towards total integration

> Advertising has dominated brand communications for decades, but its dominance is now being questioned as companies look at their communications strategies in much broader ways. (Judie Lannon)[79]

Some of the reasons why we are entering the era of integrated marketing communications (IMC) are rising costs for traditional mass marketing, lack of immediate measurable campaign effects in advertising and also a new grown insight about the gains that could be created when making it easier for marketeers to communicate their message with a single voice across multiple channels.

The integrated market perspective is, in other words, based on holistic thinking and the possibility of integrating all of the significant aspects of the marketing in an optimized mix for the company. In the midst of the rapid changes which surround us, this becomes ever more important. At the same time as available channels to the customer are increasing, it is becoming difficult to judge which of them are most cost-effective in the longer term.

As a consequence of this, the media is integrating. Television media is being extrapolated as an online environment in programes such as Channel 4's *Big Brother* (in the UK). In the case of *Big Brother* it is also possible to order SMS bulletins to keep up to date with new developments and events. The mobile marketplace is quickening the integration between various aspects of the media. Once the marketplace is truly portable, a near-perfect accessibility creates the underlying conditions for messaging services controlled by the customer, providing relevant information in any location.

More advanced software will follow up digital advertising

A lack of standardized methods for following up advertising is still holding back digital advertising, making advertisers unsure of the quality of the findings. According to a Forrester survey,[80] 46 percent of advertisers feel there is a lack of reliability and 36 percent state that they are not equipped to measure the effect of the advertising. Growing numbers of respondents withholding their names, are forcing software manufacturers to develop new authentication techniques. The SEMA Group has developed new technology to reveal false names and profiles.

Micro-geographical marketing

The evolution of micro-geographical marketing has been discussed earlier in this chapter, but to summarize and exemplify it one can say that by using highly selective and positional services such as UMTS, it is possible to reach the customer at exactly the right time and place. For example, all football aficionados within a five kilometer radius of Highbury Stadium could be sent a text message about tickets still available for a game between Arsenal and Manchester United with kick-off an hour away.

Permission marketing is becoming more widespread

The basic principle behind what is known as permission marketing is to build an interactive and mutually agreed relationship with the customer – in which the customer decides when and how much information to receive, and so on. This means that the marketer has to engage the customer in a productive dialogue – a prerequisite for this being that the marketer owns information about the specific needs of each customer.

Permission marketing is as yet in its cradle. Companies are tottering towards this new goal by starting to build more advanced customer databases; the idea being to use these systems to develop customer relationships by communicating on the customer's terms and getting to know his or her needs, so that appropriate advertising messages are delivered at opportune moments.

International address companies and postal services are working frantically to develop service models for permission marketing. Projects that have been tried include schemes to pay the customer for reading advertising messages. The principle of this is that the customer specifies his or her interests, and on the basis of this a profile is constructed. Next, the receiver is sent advertising messages that conform to this profile. An exciting, still fairly unusual version of this is that of mailround.com which offers its customers the opportunity of earning points while also livening up their e-mails with special adverts designed in the form of postage stamps. Mailround.com first collects information about the customer's interests, and then offers an arsenal of company and product stamps to choose from. The customer then receives a bonus for each stamp postmarked as sent or received.

Viral marketing

Closely related to permission marketing is something known as viral marketing. This new development in marketing is based on involving the customer in the marketing dialogue so extensively that he or she participates on a personal level, as a personal marketer to his or her own contact network of friends and colleagues. The term "viral" is inspired by the concept of a marketing message spreading almost as a bug – with the speed of a flu bug in a kindergarten. The marketing message becomes self-seeding across the digital networks, either by e-mail or increasingly via SMS text messages. In certain cases it is feasible to talk about the customer as a retailer of a particular product or service.

Viral marketing or "word of mouse" (as it is also known) is being experimented with both at local and global levels. One of its most successful exponents is Jobline, the recruitment company. Various commentators are predicting that viral marketing will integrate with developments in private resale networks. Amazon is running a promotion whereby private individuals recommend Amazon services to friends via e-mail, in return for a share of profits generated from subsequent purchases.

Among others, Levi's and Nike are experimenting with television advertising backed up by e-mail updates of the adverts distributed to customers. One of Nike's television campaigns featured a boy colliding with a man juggling chainsaws. One of the chainsaws was falling towards the boy. How was it all going to end? Soon, younger consumers started bombarding Nike's website with their ideas and preferences about what should happen next.

Public relations and marketing are merging

It is getting more and more difficult to distinguish editorial material from advertising. The marketing and PR departments of companies are more closely aligned than ever. An example of this is Pokémon, whose feature films and television programs simultaneously work as advertising vehicles for their trading cards and games. Growing numbers of people monitor the stock markets, own shares or funds, and take more interest than ever before in corporate information sources. The total information strategies of companies must therefore become more integrated – so that consumer information and marketing departments work seamlessly and harmoniously to add long-term value.

New interactive tools in public places

There are increasing numbers of interactive devices in public places. By means of various radio technologies, opportunities are being created in local environments for getting involved with information and messages disseminated within a given area – perhaps an urban area. Examples of this might include stations where information can be downloaded into Palm Pilots. It is not unlikely that we will receive messages on our mobiles or PDAs while waiting in the bus station for a bus that has been delayed. The very same company that has bought up poster sites in the bus station might be the sponsor of the message. It is also possible that within a few years, the message will be delivered as a voice message – possibly inviting us to take part in a competition in some way linked to the place where we find ourselves. A precondition, of course, is that we have given permission for advertising to be delivered to our mobile phones.

The Internet moves from being one advertising medium to many

It is no longer possible to think of the Internet as a single advertising channel. The Internet has developed into an infrastructure for a whole range of different types of communications, and hence one has to view it as a world in which a variety of tools can be used with varying degrees of success. Examples of these variations include: banner advertising, e-events, site sponsoring, advertising film for Web TV, adverts in electronic magazines or campaign websites, WAP sites, digital monitors.

The expansion in Internet advertising is going down at a controlled pace

After registering annual growth of up to 150 percent[81] growth in online advertising is now gradually falling back in many parts of the world. Monitoring organizations have stated that the current downturn is the first since 1996. The trend is expected to continue. However, there is consensus about the fact that the Internet is not finished as an advertising medium. Indications from many directions in the international arena suggest that traditional advertisers will continue to devote a growing share of their advertising budgets to online campaigns. Several advertising analysts such as London-based Zenith have predicted that 2001 will be a typical interim year – with signs of growth in 2002. By 2006, according to the Internet company Jupiter, the situation will be looking rosy once again. The industry value of the advertising market might be as high as US$26 billion, with over 1 billion people using the Internet and receiving online advertising.[82]

However, the picture becomes a little more contradictory when one looks at comments made by European advertisers in a recent survey carried out by Jupiter.

Over half of the European advertisers questioned by Jupiter expressed the opinion that the Internet was an immature medium for advertising, and hence not suitable for their purposes. More than a third thought the medium was overpriced. The value of online campaigns in terms of building brand value was still unclear. There was general dissatisfaction with the effectiveness of campaigns, existing payment models, monitoring systems and standards of service in general. On the basis of this, one must conclude that important problem areas have to be completely resolved before the market can register strong growth over the coming years:

Online advertising will not measure up to TV, print or ambient advertising until the industry solves fundamental shortcomings including customer service, efficiency and branding.[83]

SUMMARY

- A movement from product- and distribution-driven marketplaces to customer-driven and interactive marketplaces.
- Mobiles working as membranes to the digital, online environment.
- An evolution towards micro-geographical marketing with "time-and-place specific" messages and offers.
- Media convergence, free media, and increasing demands for customer utility change the media universe.
- The mobile marketplace changes the media clock.
- The rising costs of acquiring new customers, interactive advertising and greater awareness of brand attitude, change marketing as a whole.
- Marketing communication is moving towards total integration around the customer.

PART III
The Arena

We have described how new technical possibilities are clearing the way for a new mobile marketplace. We have also seen how the human and institutional preconditions are changing. The questions and insecurities are primarily about whether the technology will be good enough, and whether there will enough users to finance the investments necessary to build third and fourth generation mobile telephony.

In the following section we are going to take a look at the major players in the mobile marketplace, and how they are likely to behave over the next few years. We will also take a look at the views of some specialists, and present a study on how consumers believe they are going to make their choices.

CHAPTER 10

The Players in the Mobile Marketplace

A long line of different players are currently jockeying for position in every conceivable area in the mobile marketplace – from infrastructure companies, content suppliers, packagers and consumers. The picture is complex and the roles unclear. Some of the players have more than one role. With the trend "we're all media companies," football clubs such as Manchester United, or car manufacturers such as Volvo, become more complex entities. This makes the picture even more diffuse.

In this chapter we are going to look at the likely changes and behavior of the players in the mobile marketplace. Beginning with a general view of the players, we shall then go on to look at each one of them in detail. This will provide a solid foundation for a general analysis of the future of the mobile marketplace.

A map of the players

Figure 10.1 gives a general overview of the players involved in communication in the mobile marketplace. At one edge are companies involved in marketing or selling that wish to communicate with, and ultimately sell something to, the customer. These might be service or product-based companies, organizations, local authorities, and so on. At the other end of the system is the mobile individual who wants to communicate with the company, for instance, with the help of a mobile telephone. In the mobile marketplace the messenger has much better ways (that is, with the help of the carrier system in the middle) of reaching the customer at the right time and place with messages tailor-made to the needs of each and every

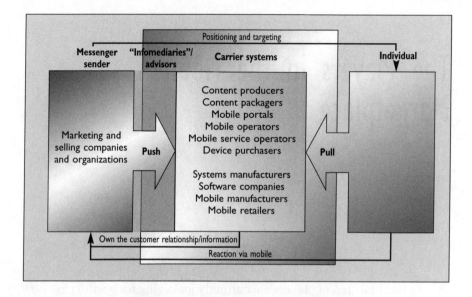

Figure 10.1 The players involved in communication in the mobile marketplace

The mobile marketplace consists of a wide range of different players with shifting roles, occasionally merging. From a communications perspective, however, they can be loosely categorized in three principal groups – senders (marketing and selling companies), vehicles (media and platforms) and receivers (people)

individual. Yet it is still unclear to what extent people are going to want such messages.

In an active context (rather than as a "victim of information") the individual has much better opportunities of being able to pick up messages, download information or order goods and services from distributors (that is, the messengers).

The "infomediaries" are the advertising agencies and other advisors that help the messengers formulate their information and choose the most appropriate channels or media.

All of the players are found in the carrier system that collectively makes up the foundation for the mobile marketplace. In this group we include content providers (either traditional media or new mobile services companies), content packaging companies (the nearest thing to newspapers or similar channels, but in digital format), portals (gathering content in a navigable form as well as providing the starting point for consumers/people using the net), and last but not least, the mobile phone network operators. We should also add all the other companies that create the new technical infrastructure such as mobile systems and other technical platforms, manufacturers of software for servers, mobile phone manufacturers – including the manufacturers of games, PDAs/palmtop computers and so on – and lastly, the mobile phone retailers.

A large proportion of the players in the delivery systems are also information mediators. The mobile service companies and content providers have to advertise their services, just as they must find ways of financing them via sponsorship, advertising and subscriptions. The mobile network operators spend astronomical sums on market communications to drum up additional traffic. It is now generally accepted that much of this market communication is not going to take the route of mobile channels. If we look at the marketing of mobile Internet players in recent history, the preferred channels seem to be outdoor advertising and other "loud and brash" types of advertising. The delivery systems will also integrate a range of new companies offering services to adapt services between different technical platforms – for instance, convergence companies offering services to help content providers scale down broadband services for use in mobile platforms. There is also an opportunity here for positioning companies, helping portals and others to make full use of positioning information in their marketing efforts.

Gathering round the customer

If we organize the map of players with value as the most important criterion, we quickly see that the customer is preeminent. For the marketing companies it is crucial to have relationships with the customer. In fact, the same applies to all players with something to sell to the final customer – companies selling products or services, content providers, mobile phone operators, mobile phone retailers. Yet, handling complex customer information might not always be their strong suit – many companies will either be unwilling or unable to do so – especially if by "customer information" we also mean positional indices. Because of this, new kinds of middlemen will emerge, such as companies specializing in mobile CRM solutions with positional data and other types of customer profiling. Internet portals are currently partially filling this role, and mobile portals will fall into the same slot – that is, functioning as an interface between "selling companies" and customers; and in addition, as aggregators of customer information. For the aforementioned middlemen it will not be actual customers that are valuable, but rather information *about* the customers. Such information will be sold on to other companies that utilize it in their own marketing. For the selling companies it is important not to abuse the customer information they collect – after all, they are dependent on the consumer's trust. Because of this, it will also be important to ensure that others do not abuse the information. The companies that will walk the razor's edge more than any others are those that have to both attract customers and act as middlemen. A perfect example of this would be the portals. The portals have to maintain constant inner dialogue about how best to proceed.

Companies, media and receivers become one

Reality is not always so one-dimensional that we can separate receivers from the media or companies. True enough, one of the clear emerging trends is that products, media and receivers are fusing – the dividing lines are becoming hazy, as illustrated in Figure 10.2. In networked games, interactive television programs and Internet communities, the receivers are actually generating the content. At the same time, more and more companies that have marketing policies are now acquiring their own channels, and their own media.

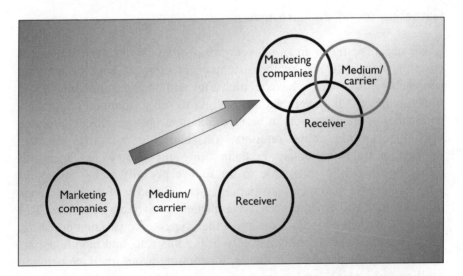

Figure 10.2 The merging of players in the mobile marketplace
In the mobile marketplace there is a tendency for players that previously were separate to merge together. Products become media in their own right, and the receivers participate in creating content

One of the aspects unique to the mobile marketplace, at least from a marketing viewpoint, is that the messenger and the receiver can be in constant contact. What this means – at least from a theoretical point of view – is that time lags disappear. The marketers or selling companies can distribute instant advertising offers, some of them valid for only 15 minutes. The receiver can answer directly. Of course, this has also been a possibility with the mobile telephone, but with text messages and positional marketing the whole process becomes much more sophisticated.

Marketing and selling companies (and organizations)

In the global, formal marketplace, retail trade goods to an approximate value of US$4500 billion are sold.[84] In the Western world, the retail trade turns over about US$4000 per person, per year.

For most of the 20th century the market was characterized by undersupply; however, over the last few decades it has become a market of oversupply both of products and services, and, one might also add, a surfeit of information. The argument could be turned around, of course. We could simply say that the market has been defined by a persistent lack of customers. This is most clear in business sectors that have been deregulated.

Another strong motivation for the changing strategies of selling companies is the post-industrial dilemma of the rising costs of employing people – hence, for instance, manual handling of customer relations is becoming more expensive.

First, there are rising demands on delivered products and services, in the form of quality, bonus services, and so on. Differences in specifications between products become minimal, and that "little bit extra" which brand value can represent becomes more and more important.

Second, the costs of gaining new customers rise as competition for new customers increases. This is arguably a simple consequence of the law of supply and demand. With customers in short supply, the price on the head of each customer goes up. At the same time, the incentives to retain existing customers are much greater.

Third, there is an increased need for mechanical handling of customers.

With the emergence of the mobile marketplace we can be more than sure that selling companies will have strong incentives to push information to customers, and to do this with maximum precision in terms of personalization, time and position. The mobile telephone offers great potential here. In addition, the companies will have a continuing need for traditional, brand-strengthening advertising, advertising that makes it possible for them to gain or consolidate their "mental market share." The goal is to "own" a relationship with the customer.

Because of this desire to develop customer relationships, it is also probable that many selling companies will develop their own portals in the form of closed systems developed in alliance, for instance, with mobile headset manufacturers or network operators. Banks, for instance, are well placed to build up broad financial portals as they are currently doing on the Internet. With this, we can conclude that selling companies are increasingly entering into the roles of channels or media. Selling companies are already broadening their identities. The British retailer Tesco is a good example of brand enlargement in various directions – petrol retailing, banking and insurance services, and so on. The Scandinavian supermarket chain ICA, part of the Dutch conglomerate Ahold, is similarly expanding its services with the recent establishment of a bank. The motivation behind this is the desire to take control of its brand – that is, keep other brands out. ICA is also planning to extend its financial services into insurance and telecommunications. It is not improbable that both Tesco and ICA/Ahold will eventually be both mobile phone operators and portal owners.

The handling and analysis of large amounts of customer data and building of systems that, based on the position and stated preferences of the customer, send tailored-made messages, is no easy task. It is therefore likely that specialists will carry out this type of service.

Advertising and media agencies, intermediaries

Advertising agencies and media intermediaries (along with PR companies) have had the task of improving the efficiency and precision of marketing communication. Their task has been to formulate the messages and choose the channels in which they are presented.

With the emergence of the interactive media, the media map has become more complicated, while the message packaging has fundamentally changed. Hence, new demands have been placed on advertising agencies, and new players such as media agencies or web agencies have appeared in the advertising business.

Consulting companies commonly aim to "get into the value chain" and reach strategic levels in their assigned companies. This tendency is also true of advertising agencies and media companies. In recent years we have seen an increasing focus on brand issues, which, in a sense, might be seen as an attempt by the advertising industry to reposition itself – the aim being to move from an industry of advertisers to one of communication strategists.

New players are also coming into this area from the world of technology – for instance, Internet consultants. While they may lack the creative communications skills that advertising agencies have built up, they compensate with technical competence in fields such as the building and handling of interactive communications platforms.

It seems doubtful if these ambitions of intermediaries – to take on strategic, advisory roles in the communications sphere – will ever be successful. The more strategic the choice of communications strategy becomes, the more likely it is to be viewed as a core area of any company. It seems inevitable that companies will seek to build up their own skill bases in areas of strategic communications, and outsourcing only on a limited and specific scale to strategy consultants, research companies, advertising agencies, media intermediaries, IT consultants, and so on. Core skill areas will be retained and built up in-house. To outsource collective responsibility for this would be like handing away responsibility for the company's overall strategy.

However, companies will probably continue to outsource the implementation of noncritical business functions. After all, this will free their hands to concentrate on strategic, critical areas. Outsourced areas would typically be maintenance of customer databases, CRM systems, and so on.

Against a backdrop of the mobile marketplace it is reasonable to expect a continued specialization among all intermediaries. The decisive factor in a more fragmented business world will be high performance in a specialized field. There will also be a growing need, not least among smaller advertisers and SMEs, for interdisciplinary advisers able to integrate functions otherwise handled by specialists. Currently, media agencies are probably best positioned to take on interdisciplinary roles. Finally, selling companies will have to put more and more energy into pinpointing long-term and sustainable communications strategies. Here also there will be a growing need for strategic consultation services.

Traditional media

Among traditional media we include one-way media such as television, radio, the daily press and factual press. These media have traditionally functioned as channels for information, entertainment and advertising. They have been grouped around a distribution channel (that is, newspaper, television, radio, film, and so on) rather than a type of content.

Also it is the media that have had contact with the customers. Customers have made their choice of daily newspaper more on the basis of their attitude to the actual newspaper than its content providers – that is, the journalists. Therefore the media company has "owned" the relationship with the customer – the only exception being the odd brilliant columnist, able to switch customers' allegiances to another newspaper.

In the interactive, digital media this whole landscape is radically changed (see Figure 10.3). First, all content is digitized (as in practice it has been for a while) so that it can be distributed in many different ways. Second, customers increasingly group themselves according to interests – this, allied to technical innovations, has led to the creation of a plethora of niche channels. Third, personal brands are becoming more and more important in helping customers make their media choices. This is especially clear in television, where the ratings war is highly developed, and where stars start their own programs, magazines and even channels.

Altogether, this leads to a modular development in which more and more media companies choose to complement their traditional distribution channels with new channels. Television programs are backed up by websites, video and interactive opportunities, or transformed into magazines on paper, or books. This means that the traditional, inviolable link between content and distribution is broken up. Consumers pick up the content of their choice via several channels. In the modular media map, the theme of the conversation links consumers, marketers, and media companies. This leads to an increased need for two-way communication.

We earlier stated that there would be two commodities in short supply in the future: the first of them, content and the second the consumer. Some types of media are better than others at generating one, or both of these commodities.

The core competence of the daily newspaper is often its local knowledge. By using this knowledge as a springboard, it can choose to act as a local agent providing information of interest both from personal and commercial points of view, and adding on mobile services also focused on the local market.

The strength of the factual press – that is, consumer magazines – is based on depth of knowledge and brand value within a particular area of expertise. In the modular media market, the factual press has the ability to distribute its content across many channels, and through this, to ultimately "own" all of the readers in each channel. To succeed at this, it will be decisive to be seen as the owner of a particular factual area. A strategy like this might eventually make the factual press a focal point for communities and other forms of interest-based collectives, both nationally and internationally. This would give rise to an opportunity of selling to these customer groups, by forging alliances with retailers. For instance, when a reader of

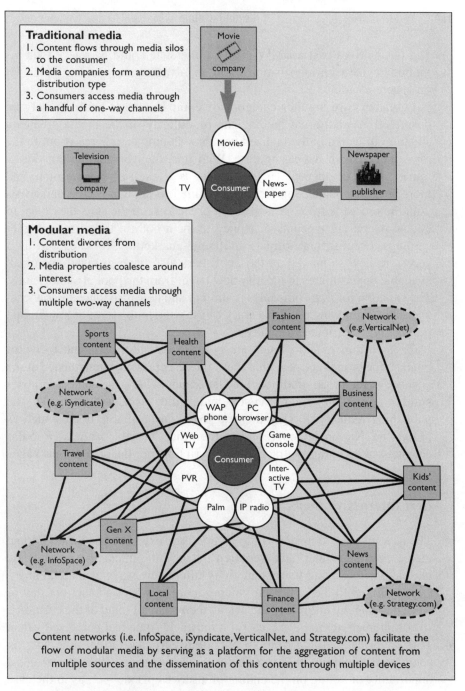

Figure 10.3 The modular media map

In the modular media map, players gather round content and customers, rather than the channels – and content providers utilize many parallel channels

Source: Forrester Research, Inc., *Deconstructing Media*, March 2000, used by permission

Top Gear drives past a BMW garage a message could be relayed via his mobile device with information about an advantageous financing plan for prospective car buyers.

Publishers are another category of companies that could loosely be described as traditional media. The modularity of the media creates a situation in which individualized brands – shining stars – reach out to the mass market with increasing ease. With this, their bargaining position is immensely strengthened. The publishers as a result of this are more and more pressed when it comes to underwriting less popular writers and artists with the proceeds from bestsellers. In order to keep the stars they have to devote increasing amounts of money to them both in the form of higher earnings, experimental formats and better marketing. To safeguard their access to new shining stars the publishers also have to make use of all available opportunities to identify new talent from the pool of submissions. The situation for film companies is similar, and here also there are low-cost alternatives allowing rookies a crack of the whip in the form of low-budget productions, distributed digitally.

Once customers and content are grouped around a particular focus, or interest, the brands active within these interest areas are strengthened at the expense of the broad, traditional media brands. This development imposes similar constraints on television and radio broadcasters as for the publishing industry, forcing them to focus on profitable brands such as MTV and National Geographic rather than broad channels. Popular channels and programs attract new subscriptions for the program package.

New interactive media

The new interactive media will to a very large extent consist of traditional media companies that have ventured out into the interactive sphere; the reasons for this being that interactivity offers new ways of maintaining a relationship with the customer, gaining more points of contact per day or week and thereby ultimately increasing their share of mind of the customer.

For interactivity to be successful it must add some real perceived value. This might be anything from an opportunity to chat in real-time with an actor from a soap, to be given choice between various alternatives or the opportunity of voting on the ending of a series/episode – or, as in the case of *Big Brother*, to be able to monitor a program round the clock.

Another form of interactivity, which has been used in the context of television commercials, is to e-mail alternative versions or endings of a particular advert to enthusiasts.

Content providers

The most prevalent trend in the overall media sphere is the rapid growth of new channels and players. Not only are new formats emerging (Internet, digital television, mobile Internet), but also there are new brands within traditional media. Examples of this are new television channels and an expanding specialized magazines press in the 1990s.

As a consequence of this, there is now a fully fledged war raging for consumer's "ears and eyeballs." During its initial establishment a new medium can be interesting purely because it is new; but in the long term it has to add something. Content, ultimately, is crucial to the success of any medium. Irrespective of whether one is in the business of writing newspapers or an Internet portal, one has to be able to deliver content (both in terms of factual quality and format) that is "sticky" – that is, compelling enough to make customers stay.

Content thereby becomes a scarce resource for the media industry and a strategic resource for content providers. The strategy adopted by players in all kinds of network structures is to take control of strategic resources. There have already been notable examples of this development in the media world. Spanish Telefonica has acquired both the portal Lycos and the Dutch company Endomol – thus gaining control in both program production and the formatting of content. Endomol is responsible for several television series that have been screened on various networks worldwide. Another high-profile example of the integration between content and distribution is the merger between America Online (AOL) and Time-Warner.

However, on the other side of the coin, the strategy of the content providers is to maximize the value of their assets by aiming for increasing independence from media, and by producing content that can be delivered across a variety of platforms (print, Internet, television, and so on). Various program production houses are redefining themselves as content providers and focusing on new interactive media alongside their more established programing output. Interactive offerings have the ability of telling a story, which is becoming so important in creating a relationship with the customer. Such narrative approaches are also becoming more common on the Internet.

With complexity increasing on all sides, brands are ever more important on the content side. Strong brands can, via digitization and New Media free themselves from the publishers and create their own channels. For instance, Stephen King delivered his last book in downloadable chapters on the Internet. However, he stopped writing the book when it became clear that few people were prepared to pay for it. Another pioneer in this context is David Bowie, who on the basis of his brand name has not only started selling his music on the net, but also set up a bank. Bowie exemplifies another possible strategy for the content provider – to create a new channel in the form of a portal. But still, only a few will succeed.

In other words, companies loosely falling within the category of publishers are paying more and more to retain their best known brands. With this, their margins start narrowing. They will not be able to finance their operations, as before, with the proceeds of a few bestsellers. The Swedish publisher Rabén & Sjögren, for instance, has financed the last 50 years of its existence with the help of Astrid Lindgren, the writer of children's books (creator of Pippi Longstockings). The same seems to be true for the publisher of J.K. Rowling's books about the young wizard, Harry Potter. A feature of the new media landscape will therefore be a very high level of investment in finding, developing and marketing new names.

In the mobile world today, it is quickly apparent that many content providers have understood the importance of delivering their content in many directions. Reuters has an alliance with Nokia and Ericsson as well as with portals such as Excite and Yahoo!, while also simultaneously creating its own mobile portal. That content providers have power is becoming more and more obvious. In November 2001 Swedish Telia, as the first European mobile operator, closed a deal with content providers which gives Telia 20 percent of the revenues from games and other content delivered over the mobile.

Content packagers

Content packagers are the closest equivalent in the digital world to publishers, and television or radio channels. They are already starting to appear in the mobile environment as aggregators of thematic information and news. An example of this is Infospace in the USA, which packages information for Internet and mobile portals as well as mobile and mobile Internet network operators. There is much evidence to suggest that we are going to see more companies of this type, while modularization continues in the media and content is aggregated thematically. The content packagers might eventually adopt an independent position vis-à-vis the customers, or act as middlemen between the content providers and the portals.

For the mobile operator and portal owners it is also complicated to handle contracts with numbers of content providers. Therefore new middlemen acting as brokers will appear in the market. As Per Nordlöt, Ericsson's GPRS director concludes: "I think there is an opportunity for brokers operating between content providers and mobile operators and portals."[85]

The portals

Not all portals are triumphal arches.[86]

In many respects, portals are the digital equivalents of the book trade and newsagents. Portals are the first stop for consumers looking for content or services. Portals, just like booksellers, own a relationship with the customer. Therein lies their principal strength and potency. Unlike booksellers, on the other hand, portals in their basic form seldom have a financial relationship with the customer. Therein lies their principal weakness.

There are many potential players in the mobile marketplace. For a comparison, see Table 10.1. Many companies are already up and running. If we view the mobile marketplace as an extension of the Internet, the presence of the Internet portals becomes a self-evident part of the equation. But to achieve any degree of success, they will have to develop

Table 10.1 Comparison between potential mobile portal players

Mobile portal player	Strengths	Weaknesses
Mobile operators	Commercial relationship (billing relationship)	Low competence in content provision
	Position information	Little experience of partnerships
	Nationwide strong brands	
Traditional portals	Substantial knowledge of the portal business	Little experience of mobile logic
	Content knowledge	No position information
	Experience of partnerships	Negligible billing relationships
Mobile manufacturers	Technical expertise	Not core business
	Contacts with developers	Not core skill area
Mobile retailers	Own customer relationships	Not core business
	Direct contact with the customer	Not core skill area
New players	Flexibility	Focused on technology
	Innovation	No experience of content development
	Focus on niches	

associations in the consumer's mind with financial transactions. Hence, portals have an inherent weakness in relation to the position of mobile phone operators.

The mobile phone operators are natural portals whose main weakness is their lack of experience in handling content and content-sharing partnerships – something that Internet portals have been doing for years and can manage "with their eyes shut."

The strength of the retailers is that they have a customer relationship of a different kind than the mobile phone operators. If much of the focus of the mobile marketplace is directed at mobile phone retailers, the latter will take on a more important role than they have today. One possible way of establishing a foothold in the market might be to offer discounts or other advantages to customers purchasing mobile phones preconfigured to the retailers' own portals.

A critical factor for mobile portals is to reduce the number of clicks necessary for any task. Studies show that with present-day slow networks and small displays, every additional click weeds out half the users.

Position players

Micro-geographical marketing looks like one of the big challenges for portals and mobile service providers. To be able to offer a good service to the mobile customer, formal relationships will have to be set up with the major content providers – which will take both time and expertise to accomplish. To ease the process, new middlemen will emerge that specialize in gathering various forms of content and handling customer data. Companies specializing in position commerce will be one of these specialized services – collecting position-related information and storing this in databases. This information can later be searched in a variety of advanced ways. These search services will also be available for portals and mobile operators, and integrated into their overall customer offering (Figure 10.4). Position commerce companies will ensure that mobile customers get information at just the time when they need it. The information will be adapted according to the stated needs of each individual, and information that is not required will be filtered out.

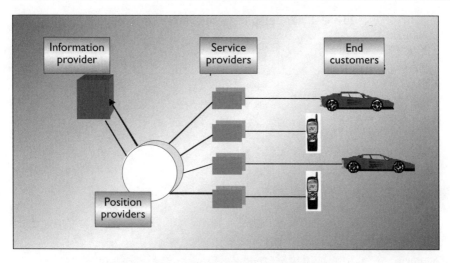

Figure 10.4 The role of position players

Position players will take up position between content providers
and service providers (portals and mobile operators)

Source: "On Position" EMBA essay, Handelshogskolan, Gothenburg, 17 May 2000

Mobile network and service operators

The mobile network operators that have thrown themselves into the scramble for the 3G networks are, to a very large extent, daughter companies of traditional fixed-line telephone operators. For a number of years these companies have subscribed to the strategy of building conglomerates – which, in plain English, means that they have sought to gather short-distance and long-distance telephony, Internet, cable television, broadband and mobile telephony under one roof. However, in recent years this strategy has looked increasingly unviable. Competition from new kinds of specialized operators is harsh. An analysis of international telecommunication shows that competitors such as WorldCom and Cable & Wireless have increased their market share from 13 to 24 percent in one year. The losers were companies such as AT&T, Deutsche Telekom and British Telecom (Table 10.2) and, at the same time, the proportion of long-distance calls made from mobiles is registering strong increases every year. A new type of operator, such as Vodafone, is also trying to take control of the global mobile market. Vodafone has networks in 29 countries and 93 million subscribers.[87]

Once the transition to 3G networks is complete, there will be two kinds of mobile operators. One will own licenses; the other will lease a share of the networks. Ultimately it is not impossible that the networks will be

Table 10.2 Telephone operators year 2000

Company	Country	Revenues 2000 (billion US$)
NTT	Japan	103
AT&T	USA	66
Verizon Communications	USA	65
SBC Communications	USA	51
WorldCom	USA	39
Deutsche Telekom	Germany	38
France Telecom	France	31
British Telecom	UK	31
Alcatel	France	29
Olivetti	Italy	28
Telefonica	Spain	26
BellSouth	USA	26
Sprint	USA	24
Vodafone	UK	22
China Telecommunications	China	21
KDDI	Japan	21
Qwest Communications	USA	17
China Mobile	China	15
BCE	Canada	15
Japan Telecom Co.	Japan	13
Carso Global Telecom	Mexico	13
Telstra	Australia	12
Royal KPN	The Netherlands	12
Cable & Wireless	UK	12

Source: Affarsvärlden 44/2000. Based on Fortune and Financial Times

separated in the same way as the electrical grid and railway network are distinct from power companies and train companies.

Richard Branson's Virgin Mobile was the first, in autumn 1999, to take out a lease on a network (Figure 10.5). Branson is widely respected as a

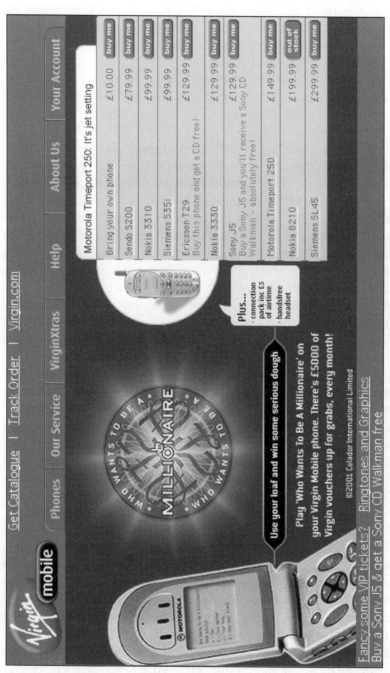

Figure 10.5 The conglomerate Virgin – with its Virgin Mobile – was the first mobile operator without its own network

Source: www.virginmobile.com

revolutionary entrepreneur, as a result of his habit of introducing new business ideas into traditional sectors. Virgin Mobile sells its phones and subscriptions at Virgin Megastores and other retailers serving the entertainment market, and capitalizes on the brand name built up among young people through Virgin Records.

The network operators are more concerned than anyone else with the ultimate success of the mobile marketplace. They have invested in licenses and network expansion. High traffic is a precondition for a reasonable return on their money.

In order to achieve high traffic, two things are necessary. It is fundamental that plenty of mobile phones are sold, and plenty of users buy network subscriptions. Hence network operators will subsidize the mobiles heavily, as they did with GSM phones. If consumers are really going to use the available services, these have to be both easily accessible and appealing. As consumers we might try something out once or twice, but unless it is genuinely useful we will quickly grow tired of it. Thus the network operators have excellent incentives for investing in user-friendly mobile portals and quality content. In view of the fact that content looks like becoming a narrow sector, the mobile operators (such as Telefonica) will take control of leading content providers. Because they are often active in many different channels, the opportunity is thereby gained to deliver content in many different areas – via cable, the Internet and mobiles.

The network operators also have a strategic resource and thus advantage (in addition to obvious advantages such as size and financial muscle) in the form of their relationships with another group of players: the customers. Direct contact with customers provides opportunities of monitoring purchasing behavior – and, in the future, mobile behavior. This will be a significant factor: in autumn 2000, the Scandinavian energy company Birka Energi pulled out of joint projects with the telephone operator Glocalnet after failing to agree on who would control customer relationships.

By gathering and owning information on the behavior of customers, network operators can in practice become content providers of customer data, a strategic information resource. To do this with any real success it is important to get the customers to carry out tasks in areas of the mobile Internet that the operators actually control – so that the information can be gathered. Hence operators are forced to choose between shutting their customers into a closed mobile intranet (sometimes known as a walled garden) or creating a portal with such compelling features that customers choose to remain there of their own free will. If the operator chooses the closed alternative, even more effort must go into creating content, as incentives for independent content providers are very much reduced as a result. Without excellent content, the network operators will "bumble along" with low traffic, hence generating insufficient revenue streams to finance their sizeable investment. In other words, the incentives are very strong indeed to generate high-quality content and services that make a genuine difference to consumers.

Mobile retailers

Mobile phone retailers have a strong position by virtue of their strong relationship with the consumer. Often they meet customers eye-to-eye. They direct individuals to particular headset models on the basis of stated preferences, and then set the phones up and get them working. It is effectively as "training organizations" that they can do their bit to speed the development of a mobile marketplace.

But the incentive to help the customers is not particularly pronounced, unless it holds some self-advantage. For as long as it is comparatively complicated to configure and make mobile phones compatible with various services, mobile phone retailers can achieve strategic advantages by selling phones preconnected to their own portals. This is precisely what companies such as Carphone Warehouse do. Carphone Warehouse is the leading European mobile phone distributor, with some 1000 stores in 15 markets. In 2000, Carphone Warehouse and AOL set up the WAP portal Mviva, available in many European countries. Because of their position as the actual suppliers of the new generation of phones required to reach the new services, they will be able to build up significant numbers of portal users. But this might also mean that they come into conflict with the network operators, who are subsidizing the equipment and who also lay claim to the customers.

The network operators could respond to the retailers' strategies in a number of ways. They could set up direct retailing chains or offer subsidies to customers opting for their closed-access systems.

System and platform providers

System and platform providers have many different constituents. The manufacturers of mobile phones, with Ericsson, Lucent, Nokia and Motorola at the head of the pack (Figure 10.6), supply the mobile networks with the physical infrastructure. But for the systems to deliver mobile Internet services, hardware and software are required to link up the mobile

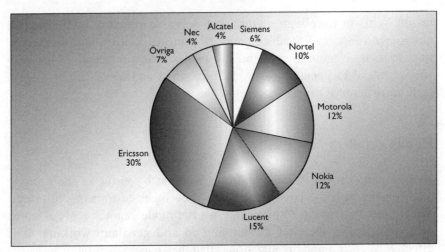

Figure 10.6 The world's biggest manufacturers of mobile systems in 1999
The total market was in the region of US$44 billion
Source: Svenska Dagbladet/Näringsliv 20 July 2000

networks with the mobile operator sites and customer sites. At the time of writing, Ericsson and Nokia and others have developed their own systems to do this, but there are compelling reasons to develop a universal standard for all. For instance, WAP forums, which are delivered by all of the leading mobile players, would benefit immensely from a universal standard.

In the category of system providers delivering more general platforms for mobile portals, mobile financial transaction services, position services, customer handling systems, and so on, there are many players and the platforms are becoming increasingly modularized. To build a functioning portal with integrated financial transaction services and sophisticated customer services, it is perfectly conceivable to buy in a number of "plug-and-play" modules and snap them together like pieces of Lego.

It is in the interest of the system providers to hype the mobile Internet in a way that stimulates both investors and users. It is also in their interest to develop efficient applications and, through this, universal and open standards.

Software companies

As mobile phones become smarter, software plays an increasingly important role. One of the questions is the location of the functions – in the mobile headset or in the network? Certain functions are suitable for use when not connected to the net, while others are only necessary while there is a connection.

Most of the impetus in the mobile Internet has so far been placed on the development of phones, and hence the networks. Much of the development has thus focused on WAP-based applications. A legion of software companies has emerged in recent years, developing everything from one-off services to general platforms for games, portals, position services, and so on. The former can be described as content providers, the latter as platform providers.

However, the picture is very different in the palmtop computer market. The operating systems for handheld devices are dominated by three systems: Psion's EPOC, Palm OS and Windows CE. Palm has a 50 percent share of the total market. As phones become smarter, the performance of operating systems and software become more critical. Over the last few years there have been several unholy alliances in areas such as operating systems, as well as new types of collaborations between the manufacturers of mobile phones and operating systems. The mobile phone and palmtop manufacturers obviously want to avoid a situation in which the software companies take a dominant market position in much the same way as Microsoft has done in the personal computer market. The mobile manufacturers Ericsson, Nokia, Motorola and Panasonic thus collaborated with Psion to form a consortium known as Symbian. The idea of Symbian is to develop an operating system for various types of mobile devices (palmtop devices and mobile phones). The interface to the user will be interchangeable, like a shell, and with this the field is wide open for all kinds of new alliances.

For the software companies (and not just Microsoft) the position of becoming a de facto standard for all of the rest seems to be just over the horizon or just below it – like a mirage. In Spring 2000, Microsoft and its development partners HP, Compaq, Casio and Symbol, launched its new version of Windows CE under the name of Pocket PC. The clear aim of the project was "to win or disappear," and Microsoft's goal is to control the whole chain from the mobile phone via the palm computer to the desktop PC.

Mobile manufacturers

Nokia dominates the mobile phone manufacturers, with more than a third of the world market, heading for 40 percent. Motorola and Ericsson, a few years ago holding about 20 percent each of the world market, have tumbled to around 10 percent each (Table 10.3).

The costs of staying in a market with shrinking margins and an increasingly competitive standoff for customers are understandably extremely high. The competition in the telephony arena in recent years is most accurately described as murderous, and consequently several companies have either bowed out altogether, or set up joint ventures in technical development and/or marketing – for instance, Ericsson and Sony, Philips and China Electronics. As the market reaches maturity it becomes less and less important for the system manufacturers to have their own models on the market.

With the transition to GPRS and UMTS, "mobile" is no longer synonymous with "telephone." Manufacturers of handheld devices have already broken into the market and will continue to strengthen their position. Compaq expects to sell more handheld devices than laptops in 2001, and as the Ipaq model was equipped with a GPRS modem in autumn 2001, this should give an additional boost to its sales figures. By 2003, Compaq believes that sales of handheld devices will outstrip the combined sales of portable and desktop PCs. And with the entrance of Russia's Cybiko, game platform producers have arrived on the scene. Cybiko's solution is based on serverless peer-to-peer technology which allows people to communicate and play network games within a range of 300 meters. It is also available with GPRS module.

The palmtop computer has its roots in a variety of designs. On the one hand there are the shrunken laptop models with keyboards, featuring word processing, calculators, and so on – such as many of Psion's models. Then there are the electronic almanacs such as the Palm Pilot (controlling about

Table 10.3 The world's largest manufacturers of mobiles, market share Jan–March 2001

Company	Market Share (%)
Nokia	35.3
Motorola	13.2
Siemens	6.9
Ericsson	6.8
Samsung	6.3
Others	31.5

Source: Gartner Dataques press release 31 May 2001

one third of the market), which use electronic pens instead of keyboards. In addition there is a variety of pure "pen on screen" computers, traditional PDAs and last but not least "smart" telephones such as Nokia's Communicator.

Once functionality is more or less indistinguishable in all the various mobiles, other factors such as design, identity and brand become more important. This has more or less already happed in the GSM sphere. Nokia, for instance, is declaring openly that it is not selling telephones but lifestyles. And Ericsson, which has been ridiculed for the clumsy designs of its phones, announced in November 2000 that it had put all design aspects of its range of mobiles under the central command of an industrial design professor; and within a few months, Ericsson announced its new joint venture with the consumer giant, Sony. The mobile is fast becoming an object of desire, a product with which people express themselves, just as they do with clothes, jewelry, wristwatches, and so on. With 3G mobiles this development will be even more in evidence; especially once the mobile gains additional functions and may be hidden in items such as jewelry, wristwatches, optical frames and clothes. In the same way as any self-respecting fashion house has its own range of glasses and wristwatches, in the future it may well produce its own range of mobiles in association with a platform manufacturer to produce the technical hardware, and software companies to provide the machines with the necessary intelligence. Both Ericsson and Motorola have announced, for instance, that from 2001 they plan to collaborate with mobile manufacturers in accordance with this model – something Ericsson has designated an "Ericsson Inside Strategy" analogous with Intel's successful strategy. Ericsson is now successfully selling its platforms to other mobile manufacturers.

The mobile individual

Mobility creates freedom. The great majority of people – not least young people – regard mobile technology as something that engenders freedom.

It is likely that mobile technology over the next few years will take on a significance not unlike that of the motorcar, which became the enduring symbol of personal freedom after the Second World War. This is nothing new. Ever since the arrival of laptops and the first mobile phones, they have been marketed as symbols of freedom. But computers and mobiles have never been hooked up effectively, so that the individual could feel truly mobile. The laptop computer is still too big and awkward – a hindrance to users who want to carry it around wherever they are. Unlike the motorcar, the laptop has not become a symbol of freedom to most people. To put it simply, it does not satisfy any real need.

With mobile broadband in the home and workplace, we will eventually expect to be able to act freely beyond home and work environments. Aspects of our behavior will change radically; just as telephone behavior has migrated from the home to pavements, buses and other public places.

A prerequisite for this, of course, is that we are allotted the necessary time to learn new behavioral patterns, time to put "new stuff" to the test. Time-pressure may be one of the foremost obstacles for the mobile marketplace, the time-pressure that ultimately stops us from trying to get that WAP phone to work. The fact that behavioral change needs time has been understood by Saab, which allows potential customers to test drive cars from the airport to their homes. In this way the company manages to borrow a little low-cost time from hard-pressed customers. What is achieved, in the end, is that Saab "owns" the customer for long enough for him/her to come away with an experience of what it means to own a Saab motorcar. In the early infancy of personal computing, a great deal of schooling was necessary to get people actually using computers. It is possible that something similar will be required before the mobile marketplace can really kick off.

People are faced with an ever-more intense flow of information. Both professionally and as individuals, people have to make more and more complex decisions. Research shows that the more complex our lives, the more intuitive individuals become. Rational decision-making is substituted for a more emotional, intuitive process – this substitution is fully conscious. Trusted information sources also become more qualitative and personal; we ask friends for advice, we seek out people whose opinions we trust.

Hence, there is much to indicate that "trusted friends" in various forms will become ever more important in our decision-making processes. These might not only be actual friends, but also public figures in which we have confidence. They might also be brands and symbols that we trust. This, in fact, is the very same role held in the medieval world by heraldic insignia, mantles or hereditary weapons.

To extend the medieval analogy further: the mobile will take on the collective qualities of the knight's horse, sword and squire. The horse helps us travel quickly and far away without growing tired. The sword sharpens our senses, helps us divide the great from the small. The squire is our personal assistant. And just as the sword was unique, personal, a cherished possession and friend – our mobile phone is also our personal, unique friend (see Figure 10.7).

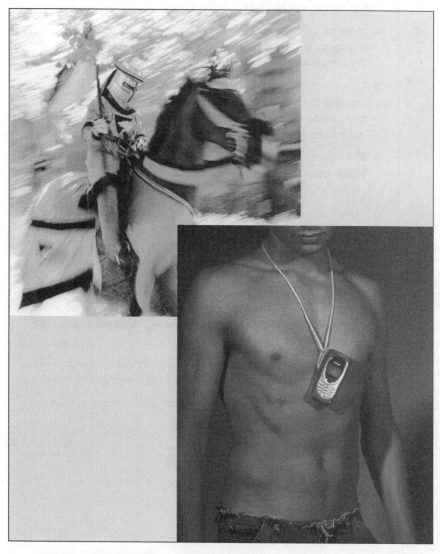

Figure 10.7 The 21st-century knight in armor is considerably more lightly dressed than his predecessor
Source: Nokia, Daniel Lagerlof/ETC Bild

SUMMARY

- A large number of players are trying to establish a presence in the mobile marketplace.
- Most of these players are either trying to "own" the relationship with the customer (that is, the business relationship) or the information about customer behavior.
- Content will be a scarce resource yet vital to kick-start demand for mobile services. Content providers will be attractive targets for takeovers. Strong content brands will strive for independence and presence across many platforms.
- The media map is being modularized; content providers will group themselves around themes and customers rather than channels.
- New kinds of companies will emerge with specializations such as position information and content packaging.
- Many new players will enter the portals business, not least the mobile phone operators.
- Handheld computers and mobile telephones will become indistinguishable; strong consumer brands will enter into the mobile phones market; and the mobile phone will become increasingly personalized as a device.
- In the face of information overload, "trusted friends" will become increasingly important in decision-making processes.
- An overview of potential interests, strategies and core assets is presented in Figure 10.8.

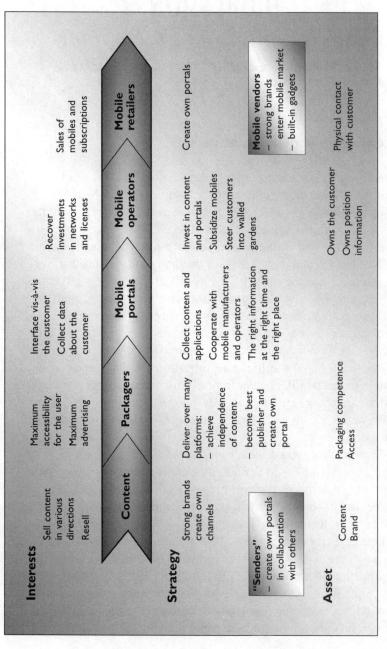

Figure 10.8 Summary of areas of interest, resources and possible strategies of various players in the mobile marketplace

DIGRESSION
Experts and Users Give their Views on Tomorrow's Mobile Marketplace

The Swedish population as users

According to the global study, World Value Survey, headed by Professor Ronald Inglehart at the University of Michigan,[88] Northern Europeans and particularly Scandinavians represent the kinds of universal values we can expect everywhere in years to come. The World Value Survey covers some 60 countries and 75% of the world's inhabitants. In the many evaluations that have been made since the early 1970s, it is clear that there is a gradual movement towards self-expression and well being; and that traditional authoritarian patterns are giving way to more rationalized and secularized views on life.

Sweden, along with the other Scandinavian countries and a few others, has one of the highest proportions of Internet-accessed people in the world. A clear majority uses the Internet regularly, and a clear majority owns mobile phones. Research also shows that Swedes, along with Americans, are the most mature Internet shoppers.

These factors would seem to indicate that a study into the aims and expectations of Scandinavians in relation to mobile technology, would give some useful directional ideas for the future. In this section, we will provide some glimpses of the hopes and fears on the Swedish horizon.

From the days when public Internet access was first available, Kairos Future was the first company to initiate continuous monitoring of the Internet habits of Swedes. This monitoring began in April 1995, when 1% of Swedes had surfing or e-mail possibilities. About a year later, when usage had gone up to 6%, we carried out a survey to ascertain what kinds of Internet service people actually wanted. Most of the respondents at that time only had a vague idea of the possibilities of the Internet, much of which had come from friends, the media or professional demonstrations. Nonetheless, the study established that people's original hopes and aspirations for the net were very much in line with how it is actually being used today. Banking services and corporate/public information services were highest on the list of desired services.

The situation with regards to the mobile Internet today is very similar to that of the fixed Internet in the mid-1990s. Many have heard of it, but very few have actually tried it out in practice. Within the framework of the project behind this book, in autumn 2000 we carried out another survey of potential users of mobile services. The idea of this was to establish predominant interests in any particular groups, access and attitudes to mobile technology and prevailing opinion on the question of releasing personal information.

Two thousand responses from a random group of Swedes aged between 16 and 79 were collected by post within the framework of the continuous monitoring project carried out by Kairos Future and FSI (The Research Group for Social and Information Studies). The response rate in these surveys is normally high, somewhere between 70 and 80 percent.

A mobile phone – and what then?

The study indicates that the mobile telephone market in Sweden is showing tendencies of saturation. Almost 78% of the total sample already own mobile phones and only 1 or 2% are still planning to buy mobiles. Of those aged between 16 and 55, 90% own mobile phones. Interest in WAP is almost nonexistent. Only a few percent of the respondents own a WAP phone, and only a few percent are planning to purchase one. The situation is very similar with PDAs.

However, ownership and planned purchases of laptop computers are at considerably higher levels. In total, about 15%

of the sample have access to a laptop computer, and in certain subgroups about half own a laptop.

A mobile generation?

The attitude to technology is somewhat contradictory. The fact of always being able to communicate with the outside world is viewed as a freedom. At the same time, constant Internet connection is seen as a straitjacket imposed by modern society. On this latter statement there was almost complete agreement. The only age group that significantly detracts from the overall pattern is the youngest in the sample, consisting of those born in the 1970s and 1980s. A lower percentage in this age group agrees with the assertion that "The necessity of always being contactable via mobile telephone and e-mail is a straitjacket for the modern individual." In other age groups in the sample, more individuals agree with the statement. Younger people seem to be more favorably disposed to the possibilities of technology. A considerably higher proportion fully agrees with the assertion that "The possibility of always being able to contact other people, wherever I am in the world, gives me a sense of freedom and independence." The differences are emphatic between the youngest age group and individuals perhaps no more than 10 years older. We may here be seeing the emergence of the first Internet and mobile phone generation.

Under the circumstances it seems significant that the youngest age group is less dubious about handing out personal information. A considerably higher proportion of the youngest age group agrees with the assertion: "I am perfectly willing to give out personal information electronically (via Internet, credit cards, and so on) to organizations of which I consider myself a customer, in order to be eligible for better prices or special offers." It might be suggested that this response only indicates that the young are naïve, but this does not appear to be the case. The youngest age group was also more skeptical about the ways in which suppliers would use their personal details. In other words, they seem aware of the risks but willing to take a chance; and they view their personal details as a commercial commodity.

Hence, there is a clear watershed in the perception of the benefits of technology between the age groups up to 30 and above. This might of course be purely down to lifecycle. Younger people are usually more open to change, and live different lives from older people. But the different attitudes might also be grounded in genuine shifts in values and lifestyles. The younger generation has grown in the constant buzz of media, with the constant themes of the Internet and mobile phones. Their formative years have been marked by technology, from the very first taste of breast milk to undergraduate's ale.

Another watershed is evident in the choices of what next-generation mobile telephony should be used for. The variables are again explainable in terms of "life phase." FSI's collections of data (built up from the 1950s and on) reiterate the obvious truth that younger people are more focused on friends, entertainment and communication. This lifecycle-oriented fact becomes significant in their choice of services. This is confirmed by a number of other studies, such as Accenture's study The Future of Wireless[89] and the Swedish Victoria Institute's survey[90] into young people's mobile habits, which concluded that the mobile was being used as a social gadget to create adhesion between groups of people.

In Figure 10.9 we can see the stated preferences of the sample for various types of services. The most popular preference, perhaps not unexpectedly, with 45% declaring their interest, is e-mail/messaging. However, 29% of those questioned have no interest at all in mobile services – the majority of this group being of the prewar generation. Of all those who express an interest in mobile services, about 65% would use the mobile to send e-mails.

E-mail is a typical mass-market service. The same can be said of other utility services such as banking, tickets and reservations, news and information surfing, or downloading information from companies and municipalities. For the record it should be stated that eight of the most popular services come under the category of

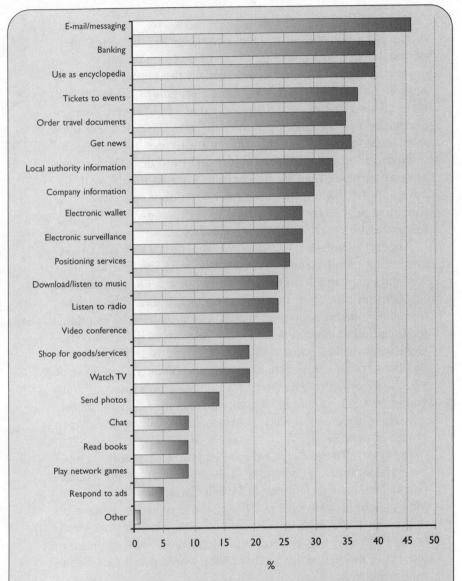

Figure 10.9 Survey: Preferences for various services of the next generation of mobile telephony

"Preparations are under way for the next generation of mobile telephony with much improved possibilities of data transfer, for example pictures. (...) What would you like to use the new technology for?" Percentages of Swedes who expressed preferences for particular services

Source: Kairos Future AB, 2001

"productivity services" (see Table 10.4 below). Narrower, more entertainment-focused services such as chat and networked games primarily attract users in an age group up to 30.

Willingness to pay

Are we prepared to pay for mobile services? If so, how? Preliminary findings in the survey seemed to give some credence to the theory that there is a willingness to pay. An overwhelming majority, 75 percent, was prepared to pay for services; however, only for services actually used, as opposed to fixed monthly subscriptions.

Key groups

Adult life begins when you have kids, people say. Friendships have to be put aside because of the needs of the family. Free time reduces to an absolute minimum. Once the children grow up and leave the roost, the somewhat aged but still vital, mature man and woman resume power over their lives. Better still, their joint income goes further without the constant and immeasurable demands of a growing family.

Age and family are hence clear watersheds when it comes to lifestyle and consumption patterns. Another strongly influential factor in most areas is education. People with a longer, more theoretical education usually have better incomes, more intellectually oriented jobs, more postmodern values, higher technical maturity, more sophisticated lifestyles; they are also disproportionately represented among "early adopters." They are what Ronald Inglehart would call "postmodern consumers."

Moklofs (mobile kids with lots of friends) are postmoderns younger than 30, usually without children and focused on their social group (Figure 10.10). They want to use the mobile to keep in touch with friends, set up meetings, chat, play games, listen to music, use video telephony and possibly also send pictures. In short, they want to use the mobile for everything. In Sweden they

Figure 10.10 Two principal groups in the mobile marketplace: Moklofs and Yupplots

Source: Nils-Johan Norenlind/Tio Foto AB, Stefan Bohlin/ETC Bild

make up 60 percent of the 16–30 demographic group. They are considerably more interested in the mobile than their nonMoklof peers. In fact, they are interested in all services except watching television, listening to music and playing games. Well-educated Moklofs are less naïve in their assessment of corporate players in the mobile market than their less educated compatriots. The Moklofs have two life projects: keeping the flock together (a desire consistent among all Moklofs whatever their education) and personal and professional development.

Yupplots (young urban professional parents with lack of time) consist mainly of well-educated and remunerated professional, urban people with families, in the 30–50 age bracket (Figure 10.10). In Sweden, they make up around 40 percent of the age group. Yupplots have extremely high expectations of themselves and what life has to offer. They want to be perfect husbands or wives, parents, colleagues, managers, sons, daughters; as well as keeping their bodies in trim and hitting their lifestyle targets. Their biggest life project seems to be to get the agenda to match up with their lives. Typically, Yupplots are mobile people – they own laptop computers, mobile phones and often PDAs too. Their primary goals are to save time, to increase personal productivity, to keep informed and abreast of change, and to utilize "dead time" effectively. They plan the week ahead on their way home from work – or similar actions such as downloading information about a travel destination, sending a few e-mails, checking news reports, making some adjustments to the share portfolio and e-mailing the week's agenda.

Sallies (senior affluent life-lovers enjoying a second spring) make up slightly less than 30 percent of the 50+ age bracket in Sweden. Sallies are middle-aged or senior people with grown-up children; they have the necessary economic clout to influence the market. Compared with other non Sally members of their demographic group, they are more interested in all kinds of mobile services. With the excesses of career stress behind them, Sallies have reached the apex of their lives and now intend to enjoy themselves to the full in the pursuit of something they truly want. The time pressure that bears down on Yupplots is not as much of a problem for Sallies. They have time to spare, with kids either flown from the nest or in their late teens. This available time they devote to golf, hunting or buying a new Harley Davidson. Their life project is "having a good time." Well-educated Sallies are much more mature about, and interested in, technology than their less educated compatriots. About 70% express an interest in the mobile Internet, as opposed to 50% of their less educated compatriots.

As we can see in Table 10.5, there are strong similarities between the groups in the designated Top-10 list of services. However, Moklofs generally attach greater importance to their chosen services. The percentages in the upper section of the table relate to the proportion of a group that has expressed an interest in a service. As we can see on the second page of Table 10.5, the percentages are 89% for Moklofs, 82% for Yupplots and 72% for Sallies. The statistics show that interest is highest among Moklofs, with 9.9% chosen services on average, against 8.5% and 7.2% for Yupplots and Sallies respectively. The most pronounced differential is for entertainment services. "The Freedom Index" is an index based on answers to two questions asked in the study – first how the respondent felt about the *possibility* of being constantly in touch with the outside world, and second what the respondent felt about the *demand* of always being contactable. The responses were interesting, showing as they did that Yupplots are the least enthusiastic about mobile technology – that is, are most stressed by the new technology. Probably this has to do with the time pressures this group constantly lives under, and the demands of family and career. "The Integrity Index" is an analogous index to the abovementioned, arrived at by asking two questions – the first relating to people's willingness to divulge personal information to various providers, the second to their confidence in the ethical behavior of those providers. It was no great surprise to find that Sallies place more emphasis on personal integrity than the other two groups. Moklofs, as we indicated earlier, are most willing to release personal details without necessarily having any real trust in the suppliers – this unlikely combination places Moklofs at the same level in the index as Yupplots, who have more trust in the suppliers but are more selective about divulging personal information.

Access to technology appears at the bottom of the table. The statistics for laptops, mobile telephones, and so on, indicate the proportion of people in each respective group that already own or plan to own a particular piece of equipment in the near future.

Here we can see, for instance, that access to laptops is considerably higher than access to WAP phones, and that Palm computers are at the same or slightly higher level than WAP phones. Access to mobile phones, on the other hand, is very nearly 100%.

Table 10.4 Three types of mobile services. Multivariate analysis shows that the services fall into three distinct groups

Productivity services	Entertainment services	Positioning services
E-mail/messaging	Download/listen to music	Positioning services
Tickets to events	Listen to radio	Electronic wallet
Order travel documents	Video conference	Electronic surveillance
Use as encyclopedia	Send photos	
Banking	Watch TV	
Get news	Chat	
Company information	Play network games	
Local authority information	Read books	
Shop for goods and services	Respond to ads	

Conclusions

The conclusions may be summarized briefly as follows:

- There is a great latent need for mobile services; most people view the mobile as something that engenders freedom.

Table 10.5 Interest for services among the three early adopter groups Moklofs, Yupplots and Sallies

Moklofs	%	Yupplots	%	Sallies	%
E-mail/messaging	88	E-mail/messaging	82	E-mail/messaging	61
Tickets to events	67	Use as encyclopedia	65	Tickets to events	60
Order travel documents	67	Banking	64	Use as encyclopedia	56
Use as encyclopedia	66	Tickets to events	60	Banking	56
Banking	62	Order travel documents	58	Order travel documents	55
Get news	60	Local authority information	57	Get news	55
Download/listen to music	58	Get news	54	Local authority information	53
Positioning services	58	Company information	52	Electronic surveillance	51
Company information	52	Electronic wallet	50	Company information	44
Listen to radio	51	Electronic surveillance	46	Electronic wallet	37
Local authority information	49	Positioning services	45	Positioning services	32
Electronic wallet	49	Shop for goods and services	34	Video conference	28
Video conference	41	Download/listen to music	33	Listen to radio	25
Electronic surveillance	37	Video conference	32	Watch TV	22

	Shop for goods and services 36	Listen to radio 30	Download/listen to music 20
	Send photos 34	Send photos 26	Shop for goods and services 18
	Watch TV 31	Watch TV 21	Send photos 17
	Chat 28	Read books 14	Read books 9
	Play network games 25	Play network games 12	Respond to ads 8
	Read books 21	Chat 11	Chat 5
	Respond to ads 12	Respond to ads 8	Play network games 3
Interested %	89	82	72
Total no. of services	9.9	8.5	7.2
No. Productivity services (of 9)	5.5	5.3	4.6
No. Entertainment services (of 9)	3	1.9	1.4
No. Positioning services (of 3)	1.4	1.4	1.2
Freedom index	0.73	0.3	0.53
Integrity index	0.98	0.83	1.66
Payment index	78	79	75
Laptop computer	34	36	39
Palm computer	13	18	14
Mobile phone	96	92	90
WAP	16	11	12
Percentage of age group	60	37	27

- The prerequisite for the success of mobile services is improved user-friendliness; present-day WAP phones are not attractive to many people.

- The most marked interest in mobile services is among younger users focused on entertainment and communication. Mobile players should therefore initially aim to capture this early-adopter market with an attractive range of services. To succeed, prices have to be reasonable with the option of micro-payments for one-off services.

- A clear majority of the most popular services are easily achievable with GPRS technology and do not demand faster data transfer speeds. Therefore there are good arguments for a speedy development as soon as functional solutions are readily available in the market.

- The question of privacy is still critical. There is a built-in skepticism about the way providers are likely to handle personal information. However, this skepticism is less marked among younger users.

- Assuming an appropriate positioning of services, much evidence seems to suggest that the consumer market will subdivide into three segments, consisting of early adopters with distinct requirements: Moklofs, Yupplots and Sallies.

SUMMARY

- A latent need for mobile services, but user-friendliness is crucial.
- Productivity services attract more or less all groups, while entertainment services primarily attract younger people.
- The privacy question is critical; few consumers trust providers to handle personal information with respect.
- Moklofs, Yupplots and Sallies are the important early-adopter groups.

Experts pronounce on the mobile marketplace – a "Delphi study"

As part of the project "Mobile Marketing for Tomorrow," a panel of experts was invited in September–October 2000 to comment on developments in the mobile marketplace. The study was organized as a sort of Delphi exercise, in which the delegates were asked to interpret how a series of questions would develop between 2000 and 2005/07. The questions were mailed to the participants; levels of expertise obviously varied from person to person and/or questions. About 60 people took part in the questionnaire.

Below we present some of the more significant conclusions, with selected participants' comments.

Views about future development

- The mobile marketplace will achieve a breakthrough (that is, reach a critical user mass) relatively quickly; a majority of respondents believe this will happen by 2005, a significant minority by 2003.

There is resistance in the form of a reluctance by users to spend money, but I believe that the patent advantages of the system will drive development on. However, the technology will not be ready for broad implementation for perhaps another five years.

- Asia is viewed as the leading mobile commerce region, closely followed by America.

There are two reason for this: first, the lack of fixed-line high-access networks; second, the lead already established today.

- The winners in the chain of mobile commerce are content providers followed by packagers (portals, and so on).

Content providers will be crucial in establishing high levels of traffic for the network operators, and will be better remunerated for their content than would otherwise have been likely. As for the network operators, they are already paying so much for their licenses that they can hardly make much money out of the deal in the medium term.

- The real money is in entertainment and diversion, at least in the mass market.

Entertainment (and pornography) have traditionally driven the development and usage of new services (film/video/Internet ...). I see no reason why it should be any different with the mobile Internet. Although it has to be conceded that mobile services are very focused on communication, and so one might be able to see another scenario emerging.

- We are going to be more proactive as consumers of information, but at the same time more selective, choosing a small number of suppliers to supply the services we need.

Eight years ago, the Internet barely existed for the man in the street. The visionaries say that the IT curve will increase at the same rate for many years yet. An emerging young generation and increasing IT maturity mean that within the foreseeable future we will be living entirely in symbiosis with a constant flow of information, which we have only just seen the beginning of.

- There will not be different mobiles for different tasks. A single mobile will be used for a range of tasks; and in the process become a personal possession.

The mobile will continue being the center for the m-Internet, PDAs will be integrated, or telephones integrated into PDAs. Accessories such as the Anota-Pen from C-Technologies preserve traditional actions within the context of mobile Internet usage.

- The intelligent home is some way off. A majority of respondents believe that 30 percent of households will have wireless networks communicating with a range of devices by 2007 to 2010, at the earliest.

I've spoken with "experts" about this. They don't believe in the predictions from industry. The predictions are there to attract investment from venture capitalists, they do not correspond with reality.

- The role of the car as a communications platform has also been exaggerated. A majority of cars will be capable of independent communication with the ability of receiving advertising and information from virtual road signs, and so on, but not until 2007 at the earliest. However, some of the experts consulted are more optimistic and look to 2005 as the breakthrough year.

I don't think advertising to cars will be a hit. Because, 1: road safety factors, 2: the car is a place for peace and quiet, 3. But advertising will come into play when we talk about positioning and navigation systems. The system knows when I have to fill the tank, and it knows that I have a Shell charge card. So it suggests I should pull in, fill up and grab a bite to eat at the local eatery ...

- A majority believes that more people will be reading virtual than printed newspapers by 2007, but there is no real consensus on this point. One in four have strong doubts and believe this will never happen.

I think that paper publications will be around until the day that screens are just as portable and ergonomic as paper (foldable LCD screens, etc.). If you're working with a computer all day, you want a bit of variation.

- Intellectual property rights are seen as an intractable problem – there is no real concept of how or when it can be resolved. More than 30 percent are of the opinion that it is impossible to resolve the issue within any existing legal or technical framework. Two answers figured most markedly among respondents: that 2003 would be a "make-or-break" year, and "never."

There is great urgency here. If this isn't resolved, the rest of my answers need to be pushed into the future by a good margin. This is THE issue.

- The mobile will be a large-scale advertising channel as from 2005, but many answers from the expert panel expressed skepticism: the advertising boom would never quite happen because unwanted advertising would be filtered out.

  It mustn't be like Spam. There have to be working applications like filters, etc. to give you choice about what you receive. When you're online you will demand as a user that if advertising is going to disturb you it has to be very relevant to you – you're not going to look at anything. If this doesn't work, the advertising market for the mobile will be dead in the water before it has even started. Everyone will get filters to exclude more or less all advertising. Let's face it, who wants to be disturbed? It's bad enough having people calling you all the time!

- Five to 10 percent of all advertising will be time-and-location-specific by the year 2005.

  This kind of technology is difficult from a technical point of view (positioning is not a problem, but there are many different commercial systems that have to be integrated). Anyway, I think the whole question of personal integrity is going to become explosive, at least in Sweden.

- A majority of the respondents believes that traditional media channels (such as newspapers, radio and outdoor advertising) will be fully integrated with online interactive channels by the year 2007. However, a sizeable minority does not believe that this development will ever take place.

  In a way you could say it's already happening, but the online and offline forms are so different that I've got my doubts about whether there is anything really meaningful going on here. Of course, a unified marketing approach will be possible across all media channels.

- It is still uncertain to what extent consumers will be willing to have directed advertising channeled via their mobiles. A slim majority makes the judgment that we will probably accept advertising, if it is personalized and if we have actually asked for it.

Peace and quiet is going to be even more important in the future. The utility of not receiving even the most personalized of messages is lower than the utility of not being bothered at all.

A lot of fun in the beginning – exhausting after six months. Should be individualized advertising messages selected according to detailed personal profiles and only on the basis of consent by the consumer.

- We are most likely to accept personalized advertising messages in shops and public events; to some extent also in queues, while traveling by public transport; possibly also in the car, but absolutely not in the boat, the home, in the workplace or while taking a walk.

At times when we're not doing anything special, when we're not responsible for a vehicle, (I mean, while we're sitting on a bus, a train etc.) or when we're already shopping, we might be willing to read advertising messages.

- Can we expect a strong reaction from the consumer against the permanently connected, "information-intensive" life? If so, will it happen in the near future, in 2005, 2007, 2010, 2015 – or never? Answers diverge strongly here. Almost half of the respondents believe that such a reaction will set in as early as 2003; nearly an equal number believe there will be no strong welling up of opinion against the new generation of mobile services.

Some groups will protest right at the start, others will do it later, and some won't do it at all. Sub-cultures will develop with their own education, values and behavior. In the longer term each of these groups will each be significant on a global scale.

Of course there are going to be "information-greens" but they will not have a major impact on development.

SUMMARY

- The mobile marketplace will start in earnest as from 2005.
- The winners will be content providers and packagers.
- Asia and the USA will hold dominant market positions.
- Entertainment and diversion have the best potential.
- The mobile will be a personal item, capable of tackling all mobile tasks.
- The mobile takes off as an advertising channel in 2005, but tailor-made advertising has an uncertain future.
- Intellectual property rights (IPR) are a major stumbling block.

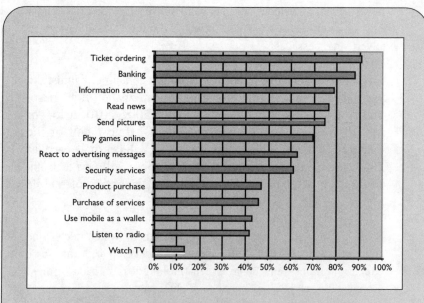

Figure 10.11 The view of experts: mobile services which will have established themselves (reached 30 percent saturation) by the year 2005

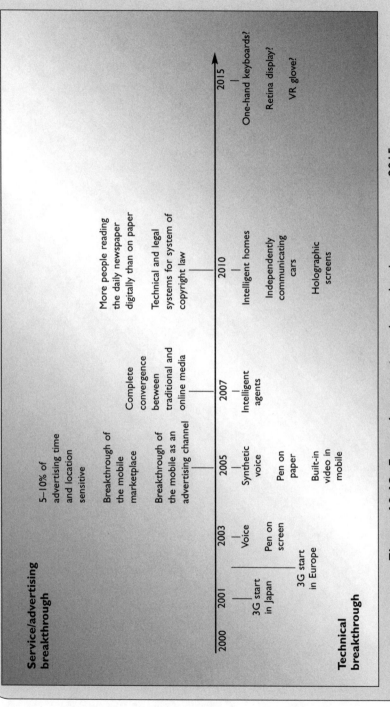

Figure 10.12 Experts' assumptions about development up to 2015

PART IV

Mobile Marketing for Tomorrow

In this fourth part we gather up the threads in the form of four possible scenarios for the future of the mobile marketplace. We start with a brief summary of the methodology of the scenario structure. Next, we summarize some developments that look like safe predictions for the next few years, as well as sketching some strategic conclusions. Once safe predictions have been looked at, we go on to the insecure elements. There are many of these – health risks, privacy, technical functionality and demand. Lastly, mindful of these insecure issues, we go on to build four scenarios for tomorrow's mobile marketplace.

CHAPTER 11

The Mobile Marketplace in 2007 – A Future Free of Surprises

One of the important questions when we come to think about the mobile marketplace and people in general is this: can anything in the future be free of surprises? The obvious answer would seem to be negative. Nonetheless, we are going to attempt to make some relatively safe assumptions up to 2005–07, and then briefly analyze the repercussions on a number of different media. First, let us look at the basic methodology.

A brief outline of the scenarios and scenario methodology[91]

When working with questions relating to the future, it is important to distinguish between three kinds of future:

- The *desired* future, often in the shape of hopes, visions or goals.
- The *conceivable* future, which might be either desirable or undesirable – encompassing threats or possibilities.
- The *likely* future, often described as prognosis, which has most influence on the imminent future.

When working with longer-term perspectives on the future – that is, outside of one's actual organization – scenario analysis is an important tool, particularly if there is a lot of insecurity or time frames are very extended.

Scenario planning is a powerful instrument for a number of reasons. First, it is an effective learning tool. Thinking in terms of scenarios helps us

Table 11.1 Examples of differences between prognoses, scenarios and visions

Prognosis	Scenario	Vision
Describes likely futures by one-off causal relationships	Describes likely and conceivable holistic pictures of the future	Describes desirable holistic pictures of the future
Points to strong connections	Well reasoned, accepts insecurities	All about meaning, direction and ambition
Obscures risks	Clarifies risks	Obscures risks, but is rooted in factual possibilities
Detailed	Gives perspective	Gives perspective
Static	Describes systems and connections	Is dynamic, and must change with the world – as we begin to enact change
Quantitative description	Qualitative description	Qualitative description
We need it to have the courage to make decisions	We need it to understand what we are making a decision about	We need it to create meaning and cohesion, but also goals and energy
We come into contact with it on a daily basis	Unusual	Common, but often empty
Functions best in the short-term perspective	Is often necessary when planning in the longer-term	Is necessary for well-being

understand developmental logic as well as distinguishing the driving forces, key factors, key players and our own possibilities of influencing events. Second, it is an effective planning instrument that helps us shape up our strategies, formulate emergency plans in the event of the unexpected, and keep a lookout in the right direction towards the relevant questions. Scenario analysis helps us develop strategies flexible enough to cope with more than *one* future.

Scenarios are memories from the future. (Professor David Ingvar)

In order to develop scenarios, distinctions have to be made between reasonably safe predictions, fundamental assumptions about the future,

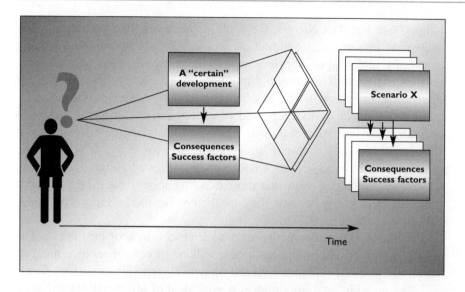

Figure 11.1 Scenario building
Scenario building is all about combining secure and insecure factors. Secure predictions form the basis of the scenario, with consequences deduced in the area being studied. More insecure developments, or uncertainties, can then be used as the bases of supplementary scenarios for the future, each with their own set of consequences

and strategic insecurities. Using the reasonably safe predictions, a basic scenario can be constructed. However, within the framework of this "safe" or "certain" development – which might be about economic transformation, increasing data transfer speeds, or mobile services – there will also be a range of insecurities. There is a limit to how much insecurity we can process – experience shows that we are only able to distinguish three or four scenarios, occasionally up to seven or eight. The trick is therefore to narrow everything down to the central insecurities, or at least look for patterns in order to reduce complexity to a manageable level. Throughout this book we have chosen to work with four scenarios, based on two main insecurities.

Safe predictions

The development towards a knowledge-based, global experience economy affects all areas of society, including competitive forces at every level. The slogan "Get Big, Get Niched, Get Out!" is increasingly accurate. In many industrial sectors mass production is a key factor, not least in mobile telephony and mobile services where the marginal cost of each new customer is negligible. Only big numbers are of interest here. To quickly gain customer mass will be a key factor in the long-term survival game.

However, in order to keep hold of customers, the ability to adapt to individual and cultural factors will become more and more important. Even the biggest and most dominant players will have to be able to adapt to the requirements of local and regional markets – as well as the global tribe. Customers are getting more and more exacting about what they want from their suppliers, partly because they are themselves increasingly working in the service sector.

The global experience economy will lead to a lack of customers. The ability to attract and retain customers will be more and more critical to survival and CRM (customer relationship management) in various forms will take on additional importance. A great deal of effort will therefore be put into cost-effective CRM solutions combining individualized, position and time-specific information enabling individualized, position and time-specific offers. Consumers on the other hand will be more aware of the value of the information they are handing out, and hence make much higher demands on what they get back.

The gigantic investments of the mobile industry players in new networks will lead to follow-on investments in what the networks will actually be used for – that is, content and services. Strong content brands therefore have a golden opportunity, and acquisitions and trading in rights will be important maneuvers in the battle for content. Strong personal brands such as best-selling writers and artists will also use their brands to create their own channels and build up packages of services for the fans.

The whole media sector is increasingly being modularized, and distribution costs are falling as a result of digitization. As a consequence, the mass channels will tend to break up and be replaced by a whole legion of channels. Rather than congregating around channels, media will to a much larger extent tend to group around interests and content.

The post-industrial dilemma will continue to whip up a productivity drive in the service industry, and heighten the need for productivity enhancement. As individuals we are pressurized to be more productive, but this pressure does not only come from the exterior world. The possibility revolution is giving us more and more choices, and the raising of personal ambition stakes means that we will increasingly strive to do more than one thing at the same time. Personal productivity and the ability to use "dead time" will be crucial.

The Mobile Marketplace in 2007 — A Future Free of Surprises

The enormous information flow will also lead to an increasing need for filtering of unwanted information. Consumers will endorse qualitative information, advice, brands and other forms of constructive intervention in the decision-making process. People will become more "rational-emotional," that is, emotional for rational reasons.

Initially mobile technology will be taken up by two consumer groups: Moklofs (mobile kids with lots of friends) and Yupplots (young urban professional parents with lack of time). The first-mentioned is community oriented and will use technology to communicate with friends, play games, and so on. Yupplots are more interested in raising personal production. In addition, companies will respond favorably to the networks, using them to keep in constant communication with their employees. After these two groups Sallies (senior affluent life lovers) will also enter the market with special needs and services.

The wireless Internet will be technically established through several, slowly integrating parallel systems. Asymmetrical radio nets and local wireless networks will emerge alongside GPRS and UMTS as alternatives or complements. Data transfer capacity in the new mobile networks will be considerably lower than promised, and probably even lower than current Internet speeds.

The mobile telephone and PDA will integrate and a rich flora of these kinds of hybrids will be established. The mobile phone will also embody much more of one's personal attributes, and hence become an object through which personality is defined.

Communication services such as e-mail and chat will establish quick bridgeheads. However, the profit margins are low in these areas. Just as on the Internet it is difficult to charge for low-value information such as news and rudimentary stock market quotes. Services of this nature will be paid for by sponsorship. The willingness to pay is highest in areas such as networked games, betting, entertainment and share dealing. Historical precedents (directory enquiries, for instance) will make it easier to charge pin money for some kinds of service.

The mobile marketplace will function as an advertising channel for direct sales promotions (such as, for instance, a timely suggestion for a lunch restaurant relayed to people who are actually on their way to have lunch), or an information source for customers that are actually looking for information. But the mobile is a personal object, and tolerance levels vis-à-vis anything that is perceived as intrusive will be extremely low.

The successful mobile portals will work within the limitations that apply in the mobile marketplace – that is, limited bandwidth, small displays, and the fact that the duration of uploaded time is likely to be low and hence not

Table 11.2 Summary of general assumptions of the coming years

Surrounding world	Technology
Knowledge-based, global and hyper-competitive "experience" economy	Technology is not a narrow sector, but bandwidth is
Sophisticated, technically competent, convenient, rational-emotional individuals	Alternatives to mobile networks – DAB, wireless local networks
Possibility and information explosion, time efficiency and minimization of decision-making	
The economy and players in it	**Receivers and purchasers**
Development driven by players – pay back time!	Decisions deferred, possibilities kept open
Digitization, modularization, blurring of divisions between media, new alliances	Functionality and micro-payments are prerequisites
Hunt for content, rights, and acquisitions	Empty channels break through first
	Willingness to pay tied to payment habits and sense of service (games, entertainment)

suitable for browsing. Hence the information will need to be condensed, with quick search routes and effective presentation.

Small, new players will establish presence in technology and service delivery, and some of them will also become operators. Large, established players from other sectors will also get involved, using their customer base to take a share of the mobile marketplace. The industry will quickly consolidate and only a small number of players will survive the first few years.

Consequences for media agencies

Advertising and market communications have historically been viewed as entities that are separate from the core activities of companies. But because of greater emphasis on brand value and other immaterial values, the prospects for core activities have themselves changed radically in many sectors (see Chapter 9).

An increasing focus on the strategic significance of market communication and a more complicated channel structure will make the concept of media choice much larger, broader and more extensive than before. Continually rising complexity in the media sector will put pressure on the expertise of the media agencies' consultancy and advisory services.

It is absolutely feasible that new players may challenge the supremacy of the media agencies in these important areas. PR agencies, for instance, are moving close to the borderline between strategic information and marketing. The ever expanding networks of the advertising agencies are incorporating the services of so-called search consultants whose job it is to put together teams qualified to handle and deliver strategic marketing competence. Even traditional strategic consultants seem on the verge of moving into the sector.

Consequences for marketing companies

The digital arena for all possible communication channels has only just started opening up. Traditional mass marketing is going to be reappraised in the light of new possibilities in developing customer relations; and relationship marketing will become much more important than it is today. M-commerce will develop in a number of directions (from the marketing perspective). We emphasize four of them:

- Constant Internet connection.
- Increasing focus on customer-led solutions.
- The mobile marketplace modifies the marketing process.
- Marketing and sales organizations become more effective with the help of mobile developments.

Constant Internet connection

Developments in the mobile world are ushering in a new concept for marketers to respond to – constant Internet connection. This is an idea, and a dream, that has been around ever since the Internet was established in earnest. In those days, however, the annoying little cable needed to connect the computer to the telephone line hampered constant connection. The mobile marketplace is on the verge of seriously creating the Omninet that for so long has been nothing but a mirage on the horizon – a natural continuation of the Internet, intranet and extranet. One important difference is that a constant Internet connection via the mobile immediately reveals the identity of the user. On the Internet, the only available information is the Internet protocol (IP) number, which represents the customer's computer – thus providing some anonymity. The positive interpretation of this, from the marketer's point of view, is that it becomes possible to carefully monitor the consumer throughout the purchasing process, and thereby gain an idea of his or her mental state. On the basis of this, communications can be personalized. The negative interpretation is that customers might choose to become more restrictive, and refuse to receive any advertising messages for fear of exposing themselves.

Consequences for marketers

Marketers are going to have to develop their sensitivities to nuances; and avoid any encroachment on the integrity of individuals. The ways in which such encroachments might happen increase with rising numbers of dimensions at play in the personalization of mass-market communication.

Successful market communications will demand more customer support through the entire process (both technical and manual – that is, human).

Increasing focus on customer-led solutions

Companies will become more intent on owning customer relationships and knowledge about customers. In the midst of an increasing complexity of channels, customer profiles will become the hardware that everyone – that is, media players and marketing companies – are looking to own.

Nor will it be enough to simply know the customer. In the experience-based economy now emerging, the crucial thing is to develop and convey one's immaterial values and thereby create one's own niche in the consumer's mind. The only financial transaction that will occur in the future is the commerce in human attention.

This work will first and foremost demand the development of a strategic customer-led approach – by choosing the customer as the strategy. This leadership of the customer has to be defended by sincere initiatives, and the relationship with the customer deepened by means of dynamic and intelligent dialogue (Figure 11.2). Rising advertising costs have resulted in larger numbers of companies seeing themselves as media in their own right, and thus setting up their own media channels for their customers. In this way, savings can be made in traditional media approaches, and in their place more refined approaches explored in owned networks and channels.

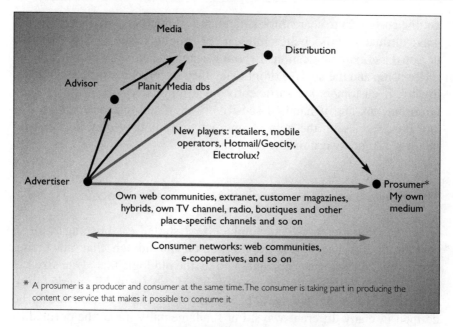

Figure 11.2 Choosing the customer as the strategy
Customer leadership is about choosing the customer as the strategy, and developing loose relations and direct contact with the customer

Develop trust management

> To gain by the loyalty of customers, you must first gain their trust. That's always been the case, but (...) where business is conducted at a distance and risks and uncertainties are magnified, it's truer than ever. (Frederick F Reichheld and Phil Schefter)[92]

The quest for customer relationships is closely related to the fact that competition is driving companies to attach greater value to loyalty and brand loyalty. Herein lies the key to long-term profitability. The abiding challenge is to achieve this in the mobile marketplace.

To develop the process of building customer loyalty, and get the customer's consent for advertising in the new mobile media, more research needs to be done on customers in various mobile situations. What it's all about is being known to be trustworthy, and thus getting customers to share their information. Frederick F Reichheld and Phil Schefter wrote in the summer edition of *Harvard Business Review* 2000 that e-loyalty is about getting customers to share information until an intelligent dialogue is under way; ultimately this creates interest for products and services through personalized communication. Learning about customers' attitudes to the relationship and the degrees of intimacy they are prepared to accept creates loyalty. If customers are sufficiently loyal they may have lower privacy requirements, and if handled sensitively they may ultimately turn into ambassadors – with the help of interactive tools they may thus end up putting in a good word for the company.

The mobile marketplace modifies the marketing process

Ever since the electronic marketplace first started establishing itself in earnest, it has been plain to marketers and admen that marketing strategies can no longer be formulated by the old logic that "One size fits all." The new logic creates new possibilities; and the mobile marketplace is ushering in further changes. The marketing process can obviously be defined in many different ways, but fundamentally it may be defined as *being precise about processes to create contact, customers and relationships*. Every time there is a new technology shift, the process is modified to varying degrees in response to new possibilities as well as threats to the status quo.

Market surveys and product development may become more effective as a result of developments in mobile commerce. All marketing is based on insights into how target groups reason and function. Traditional, quantitative marketing research methodologies can be slow, costly and static. It is no longer acceptable to merely carve up the market into structured segments. Marketers have to take on board the fact that customers change more and more rapidly in their behavior, roles and self-expression. No customer in the whole world would recognize himself as belonging in the aggregated segments put together by companies! As the mobile marketplace expands – quite independently of segmentation – marketers have to identify and separate (in addition to their work with CRM) various dynamic decision-making processes, whether or not these coincide with their existing ideas on segmentation. New marketing research methodology is therefore based on qualitative methods – allowing behavioral experts to discover deeper consumer perspectives.

The mobile is already a widely used tool for keeping in touch with participants in surveys. With the additional help of inbuilt video cameras, SMS and other methods, the opportunities of doing qualitative research will improve even more.

The Mobile Marketplace in 2007 – A Future Free of Surprises

Mobile Opinion is a Scandinavian company that has attracted a fair deal of attention on account of its fresh approach. The company maintains an international panel of mobile-equipped consumers who, within a matter of minutes, can provide answers to questions from marketing companies.

The advertising of the future has to be formed with due consideration for the fact that platforms will place increasingly high demands on market planning and stress the importance of recovering their advertising costs. This will result in more and more following up and monitoring of the effectiveness of campaigns. It will become more important to ask what kinds of multichannel strategies are relevant in order to cover the key requirements that the company can and wants to satisfy in various stages of customer development. Further on in the chapter we will discuss the possibilities of different advertising media, as well as threats to the mobile marketplace.

Contact-generating processes

Contact-generating processes are various types of marketing and advertising approaches aiming to arouse interest and create reactions and new customer contacts in the marketplace. This might be through stimulating greater awareness of the company, developing image or building and nurturing the brand.

The interactive media are on the verge of a whole new development, in which it will be possible for companies to develop customized communication. Digital television and other two-way media will enable a detailed and rapid follow-up of each individual's response, thus ultimately creating one-to-one communication.

Mobile advertising might in the same way offer a refined bridge between traditional one-way media and the customer. To this one might also add that the mobile will make it easier to extend traditional advertising campaigns. Campaigns that first feature on television and then move into traditional websites can be given an additional lease of life in mobile portals – this is also another way of improving the cost-effectiveness of campaigns.

Spontaneous reactions become possible in the mobile environment, and the mobile thus becomes an interactive add-on to traditional campaigns. Even today, consumers can use their WAP and GPRS phones to find out, for instance, what sort of jumper a television presenter on a particular channel is wearing, how much it costs and where it can be bought. Even today, one can scan in featured codes in classified ads and thus respond directly to the message. A similar approach also works very well with music played on some radio stations. All one has to do is check the mobile portal of the radio station and order the CD to be delivered to one's home.

Because most transactions are local and the mobile is the only instrument that can realistically handle position-critical information, personal advertising via the mobile will also become widespread. We have already mentioned micro-geographical marketing in this context. The implication of this, of course, is that when a user walks past a cinema with 10 empty seats for the next screening, he or she will receive an offer to buy tickets at half price. Another typical example might be an alert beamed out to drivers along a certain stretch of road that there is an Exxon filling station coming up – do they need petrol? In Sweden and internationally there are already experiments with so-called e-streets where every shop sends micro-geographical ads to people walking down the street. The results of the Swedish experiment are encouraging and sales in shops have increased.

However, there is a fair amount of uncertainty about how much advertising people are prepared to receive via their mobiles.

Some commentators are very ebullient and believe that advertising will be the savior of the mobile portals. Mediatude subscribes to this point of view, making reference to a study in which two of three users thought that advertising was a good way of getting free SMS services to the mobile. More and more advertisers are taking an interest in SMS technology as a channel, particularly for reaching younger mobile users.

European studies show that young people are the predominant user group for text messages. Examples of advertisers that are already holding SMS trials as part of their marketing include McDonald's, Maybelline and Budweiser. The research company Harvey Research carried out a detailed study into the mechanics of advertising during the European Championships in football in 2000, and specifically, traffic to the information site www.quois.com. By registering with this service, users received SMS text messages of any updates in matches during play. The SMS messages took the following form: "Denmark – France 1–0: 13 min: Thomasson. Brought to you by *Sony Playstation*." The spontaneous brand recognition was found to be around 60%, and 10–20% of the users took the opportunity of visiting web addresses included with the results.[93]

The advertising of the future will not be built on sending out advertising messages that disturb people in the middle of whatever they are doing. Rather it will be advertising by consent – making exchanges or interesting users in attractive deals – with the idea of creating dialogue. The more customized the advertising is, the less it will be perceived as intrusive; in fact, more as a source of useful or relevant information.

The Mobile Marketplace in 2007 – A Future Free of Surprises

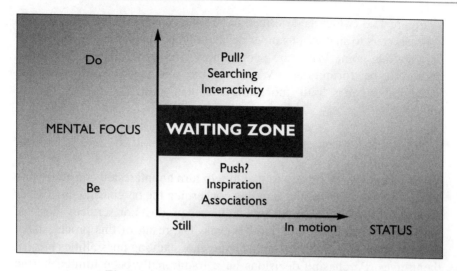

Figure 11.3 Mental states of consumers
Our mental state of mind determines how receptive we are to advertising

Other commentators are much more pessimistic and state quite categorically that people's hurried lifestyles cannot be reconciled with the idea of a constantly beeping mobile bombarding them with advertising messages. Rather, they believe the mobile is seen as nothing more than an effective gadget for quick messages.

Regardless of which point of view is correct, there are three limiting factors that have to be recognized: *the size of the target group* that mobile advertising can reach, the *transfer speed* and *for how long the mobile is used*. According to Forrester, the total time available for advertising is only 20 minutes per month.[94] In Figure 11.3 we have illustrated some mental states of mind of consumers from the point of view of mobility. There will probably be big variables in the sorts of information that will be accepted or even appreciated. Consumers might conceivably be more receptive to push-communication on the condition that the information is expected or a part of an already accepted service. In another situation, information-seeking is the relevant factor – that is, consumers actually want to get more deeply involved in something because they are actually looking for information. The problem for marketers is actually knowing when consumers are receptive to advertising communication. Tomorrow's mobile equipment, with permanent Internet connection, will most likely notify marketers of the times when consumers are prepared to accept advertising.

Figure 11.3 (which is described in Chapter 3) should be used as a basis for marketers to analyze customer behavior with mobility and mental mode as a starting point. When are people most receptive to advertising? How should it be formulated? What are the alternative communication strategies? When is "pull" more advisable than "push?"

Customer-generating processes

By this we mean all processes that aim to turn an interested consumer into a customer; in other words, anything concerning the purchasing process, or anything that identifies needs or establishes products or services that best meets the customer's exact requirements. The result of this process is, at best, an order. It is at this point that many new exciting possibilities present themselves. Purchasing decisions have traditionally been limited by the actual geographical location of the marketplace – that is, the fact that the selling channel was limited by place. But with the mobile phone, information that directly precipitates a sale will be available wherever the consumer is – whether in the high street or in other environments. During this phase it is likely that the marketing companies will own or control the media channel (as opposed to the seller), which creates a sense of security.

The payment and delivery processes include all of the processes that have to work for the seller to receive payment and the customer his or her product/service. A great deal of work remains to be done both in terms of improving the payment systems and developing new, creative ways of delivering the item, whether physical or digital.

Relationship-enhancing processes

By this we mean processes concerned with everything that happens between the buyer and seller after the delivery and payment have been made. It is also about helping the customer get maximum use out of the purchase, providing after-sales service and building long-term relations with the customer.

Mobile data communications obviously have big advantages in any context where the customer is moving around, particularly if the product/service is most useful while actually moving around. Yet again, this is all about making full use of mobile services to create real difference for the customer.

Mobile (m) advertising might also help make a particular kind of advertising more effective, namely "viral marketing" or "word of mouse" advertising. This is often about encouraging loyal ambassadors to spread a positive message about the company, perhaps via SMS or Instant messaging.

The purchasing cycle and m-communication

Below are the main phases of the purchasing cycle, and how the mobile will be used in respective phases:

- *Signal phase:* Advertising communication is there to strengthen a weakness, or a created weakness. M-advertising complements traditional advertising. Consumers, according to research, appreciate the television interface that is appearing everywhere. Public appetite for competitions is high, which can be used to strengthen brands. These create basic confidence in the brand.
- *Information search phase:* Use waiting time (that is, when people are killing time) associated with movement to go deeper into information. Permission marketing activities. Specify requirements through customized dialogue. Packaging of offers. Education of customer. References and communities/chat. Informative dialogue, tests, and so on.
- *Values/decision-making phase:* Enable comparisons between different suppliers. Price-related tailor-made offers in real-time: discount vouchers, micro-geographical context.
- *Confirmation phase:* Adapted best-in-test information, product education; tribal culture becomes more dense, changed and clarified by m-commerce.
- *Repurchases phase:* Measures to reward loyalty reinforced by interactive dialogue via mobile.

Marketing and sales organizations become more effective with the help of mobile developments

Developments in the mobile market can give companies excellent new ways of improving their information structure so that they can respond more quickly to existing market conditions. This is relevant to both internal structures (ERM – effective resource management) and customer-oriented structures (CRM – customer relationship management).

Some companies can achieve clear advantages as a result of this development; for instance:

- *Sales-driven* companies, that is, companies with a relatively high proportion of sales personnel. The advantages are, for instance, the possibility of offering real-time prices, quicker sales offers, more personalized information and immediate response.
- *Service-driven* companies can develop advantages by purchasing information with high positional precision. The mobile can help reduce costs per transaction and offer immediate response.
- *Logistics-driven* companies, for instance taxi, courier post, airlines, and so on. New business opportunities through new advanced logistical solutions. Can also monitor exact geographical position of customers, cars and resources.

Traditional media and the mobile marketplace – possibilities and threats

Table 11.3 presents an analysis of conceivable possibilities and threats for some important advertising media in the mobile marketplace of the future. We have also suggested conceivable future prospects for other media in the light of these possibilities and threats.

Table 11.3 Traditional media and the mobile marketplace

Media	Possibilities	Threats	Prospects *****
Internet media Banner advertising, e-events, site sponsorship, advertising film in web TV, ad spots in web radio, adverts in webzines, campaign sites, digital notice boards	Regardless of time and place, consumers can gain access to detailed information. Reactions in the whole chain can be monitored Geographical proximity results in direction services projected at customers	High contact costs – according to media bureaux Lack of standardization for monitoring services of Internet and mobile advertising Do portal constructors and web marketers understand the importance of utilizing the mobile to the full?	****
The physical marketplace Boutiques, shopping centers, high street	Possibility of further conceptual reinforcement and brand communication close to the actual purchase The redefining of the retailer from merely a retail channel to a marketing channel. Positioning – registering of anonymous customer streams More effective sales-generating advertising. Sending of customized messages/advertising offers Integration of virtual and physical marketplace Extension of the purchasing process to before and after the physical visit	Suppliers increase power by direct communication with customers even on the ground of the distributors The consumer's ability to compare the retailer's prices and offers with geographical or virtual competitors Unintentional showroom effect Traditional thinking	****

cont'd

***** Indicate how successful we think that the media will be in the mobile marketplace

Table 11.3 cont'd

Media	Possibilities	Threats	Prospects ******
Event marketing Events, experiences, sales fairs	Positioning creates more effective contacts and contact points Registration of customer streams Advantage of mobility during events Extension/facilitation of events Virtual fair coupled with geographical fair	Unmanageable amounts of information to handle Suspicion about excessive levels of registration Traditional thinking	***
Outdoor advertising Traditional poster advertising, digital outdoor advertising	Tie offline and online together Activate a silent mass medium. Send place-specific advertising to passers-by More nuances in the content and more rapidly switching messages (possibly depending on the time of day) Foster spontaneous reactions, engagement or shopping Prolong the effect via the mobile	Might be perceived as intrusive Environmental weaknesses (aesthetic) of the medium The market for digital outdoor advertising still in its cradle Increasing resistance to visual pollution, talk of mobile-free zones in cities	*
Print media Daily press	Use local knowledge Offer local guides with positioning and place-specific services	How to charge for the information The threat of being made out-of-date by introduction of plasma screens	***

Popular press	Consumers used to the fact that news and advertising presented together in newspaper format	Change in business logic. The willingness to pay for information is nearing zero
	Interesting content can be minimized for mobile use	
	Subscription experience	
	Payment systems	
	Certain amount of WAP experience	
	Strong, purposeful brand	
	Content that is interesting and suitable for mobile. For instance, dating and horoscopes. Experts at broad lifestyle target groups	Competition from portals expected to be significant in this area
		Increasing competition from niche communities
Specialist press	Deepen interest in most relevant locations – that is, in car, about the car. Offer advertisers opportunity of reaching customer while customer is physically active in their professional field	Digital publications, fragmentation and so on
		Increasing competition from niche communities

cont'd

Table 11.3 cont'd

Media	Possibilities	Threats	Prospects ******
Direct marketing (DM)	The next natural step of Permission Marketing	No experience of mobile logic	**
	Segmentation, CRM knowledge and loyalty will be in demand in the mobile marketplace	Lacks content	
Direct advertising	Instant reply. Sharpen DM by automating response	Traditional direct advertising is going too well for organizations to make a serious play for CRM	
	Possibility of staying with customer longer during the purchasing process	Automation in marketing communication is inevitable	
	Better at reaching customers in their rapid identity switches	Skepticism among the young to direct marketing is on the increase	
	Catch the more DM-allergic younger consumers in their natural environment: mobile, interactive		
Telemarketing (TM)	More spontaneous reactions	A certain aversion to the intrusive effect of TM	
	Increasing importance of call centers	Tougher legislation	
	Automated reply and ring-back. Speech synthesis makes many new services possible. Possibly, the TM company's information agent might contact the customer's information agent for a machine-to-machine TM conversation	The development of information agents relates to the evolutionary race – every automatic sales service immediately gains a counter-service	

Ether media		
Traditional digital TV	Unique possibility for activation and interactivity via reply-back channel	Remaining sense of interruption
	Build up chat communities with actors in soaps	Channel overload, content paucity
Web TV	More mobile productions	Traditional thinking
Radio	Prolonged campaigns	Traditional formats still rather effective
	Repackaging of content for mobile environment. Download short sequences of already-featured material	
	Shopping channel with improved prospects	
	Opportunities for product placement and spontaneous reactions via mobiles. For example, information about a presenter's blouse, and ordering it directly	
	RADIO:	
	Radio is naturally mobile	
	Hardware manufacturers making plans for integrated radio, MP3, telephone, and so on	
	Radio better suited to mobile environment, as consumers often do many things at same time, and certain activities (such as driving a car) require undivided attention	***

Scenario for the future – a user family in 2007

Let's take a trip into the future, make an imaginary visit into the home of an ordinary family living in Chelsea, London, in 2007, and see how they live – working by our designated "safe predictions."

On first inspection, all may seem perfectly normal as we enter the home of the Memosa family. But as we take a closer look we begin to realize that many aspects of their lives are radically different from our own.

To begin with, the structure of the family is not like a normal family that we have ever seen. The house is partitioned into three more or less separate parts: the largest section for the use of the core family, that is, father Max, mother Therese and youngest son Alex; a second section for the grandfather, who moved in when he got divorced from his wife at the age of 76; and the third section, exclusively inhabited by chaotic teenage daughter Tina. The IT systems of the house are based on Bluetooth technology. This makes life easier – all the different kitchen appliances are capable of communicating with each other and with the external world. Quick-dial numbers on the various mobile devices carried by family members maintain communication between all the different entities in the home.

This is what one might describe as a mobile family. Max works as a consultant; although most of his assignments are concentrated in the Chelsea area, he does a fair amount of travel. Much of the dead time syndrome he previously experienced while traveling has progressively been weeded out. Nowadays he is able to do much of his work or personal tasks while moving from "A" to "B." Max is interested in new technology, and buys early versions of new mobile devices whenever available. He is looking for ways of using such devices to make his day-to-day life easier. This year, 2007, he has invested in a Filofax computer. The name Filofax reminds Max of the good old days in the 1990s, before the Internet came along. The new unit looks like an improved version of the old Filofax, it even has the same smell. It has a leather case with pockets, buttons, leather clasp and pad, just like the original. You write with pen on paper but the information is stored in the little computer, later to be sent as e-mail or refined in the computer. The device is equipped with a little screen for GPS navigation and a location finder. Above the screen is a very small video camera, a microphone, a speaker and earplugs. The monitor is placed just above the pad, so you can see the person you are talking to. The screen can easily be folded down, as it has been found that people often only want to look at each other during the initial greeting, or possibly if something visual needs to be demonstrated during the conversation. At other times it is more expedient to concentrate on communication based on text or sound.

The Mobile Marketplace in 2007 — A Future Free of Surprises

Max likes to joke that the new and smaller technology makes the good old home office nowadays seem overdimensioned, adapted to the days of chunky computers and monitors. All that old technology seems like it belongs in another lifetime.

With a smug grin he gets into the car and starts on the journey home, having spent the whole day with a client in Manchester. It'll be good to get home, he says, mentally steeling himself for a long and slow journey. He activates the autopilot to kick in at speeds of less than 20 km/hour – that is, whenever he gets stuck in traffic jams. Once he's on the road, the car's integrated computer system asks him if he wants to go through the latest news bulletins on a shopping channel he subscribes to, or open his mailbox. He opts for the news bulletins. Immediately, the onboard navigator screen between the seats changes to a news layout with up-to-date share prices and sports stories. Once these have come to an end, Max is asked if he is interested in an interesting message from the toy retailer Toys R Us. Max has done some consultancy work for the company. Following a large purchase he made on the occasion of his son's birthday, he was placed in a special register of customers eligible for special deals. Toys R Us offers to upgrade a computer game, which, it was noted, his son was recently playing in the car. Would Max like Toys R Us to upgrade the in-car game? If so, the company will give Max 150 loyalty points and the opportunity of buying any product of his choice at half price next time he visits the site.

To begin with, Max is annoyed at this invasive suggestion. But then his thoughts start to wander. He thinks about the time when his five-year-old Alex first sent him an Instant Mail. He really felt like a proud father that day! The message consisted of a drawing Alex had made in kindergarten of himself and Max in the sailing boat last summer. This is the kind of pleasant surprise that has made Max favorably disposed to the new mobile technology.

So, with a smile on his face, he replies to Toys R Us that he would welcome an upgrade of the computer game.

Of all the Memosa family members, Tina is most interested in using her mobile – a newly purchased Westlife model. She is proud of it for several reasons – partly because she designed it herself to the acclaim of her friends, and partly because she paid for it with her own money. Its ring-tone is a languorous Westlife ballad. On the grip a scanned-in picture of one of the guys – her personal favorite – lights up like a neon sign whenever the phone rings. She has chosen a screen size that makes the phone perfect for playing her favorite games. She plays these whether she's at home, in town, on the bus, or in the car. Tina has become a bit of an expert at one of the games; each new episode is mailed to her in ascending sequences as she becomes more advanced. Every time she reaches a higher level, she is offered a rebate in one of the many virtual reality/experience centers that have mushroomed in recent years.

Suddenly Tina hears the sweet sound of the boy band singing. She looks around, but the phone is out of reach, so she answers just by saying "hello" and thus switching on the voice-activated loudspeaker. A sound mail and postcard appear on the screen – a friend's face pops up, a wall of thumping music all around her. With shrieking voice she says that Tina simply *has* to come down to the Experience Center straight away. They are playing a really funny mobile war game and there's a rumor going round that Tina is going out with the only half-decent Westlife copy in the whole school. Is that true? She's going to just *die* if Tina doesn't tell her the truth! Tina checks her GPS and is hence able to confirm that another friend (Eva) is also on her way to the Experience Center (maybe for the same reason?). She checks to see if any other friends are going there.

On her way down to the center she hears a new song by her favorites on the freestyle radio. Quick as a flash she grabs her mobile, checks to see what the song is called and the channel where it's playing. She downloads the song by pressing the Memory function. Five pence is added to her phone bill.

The family's electronic mailbox on its private intranet recently had an offer from an insurance company. It was an advertising message strictly in line with their instructions about the sort of information they were prepared to receive.

The message was dispatched to Therese's mobile filter, which now dispenses the message at a suitable moment when she's looking for diversion – perhaps while on her way to somewhere, or while standing in a queue. As it happens, she is actually queuing at the pharmacy when the message comes through. The offer is for a piece of software enabling improved monitoring of her father's health. The program is connected to a health site maintained by a local private hospital. Using sensors connected to the mobile, her father can now continually check his heartbeat in real-time.

As it turns out, her father is not especially thrilled about this. It makes him feel old and frail, and he is always "accidentally" forgetting to attach the sensors at the right time, and in the right place. Therese has made sure that the system alerts her when this happens; at such time she contacts her father IVM ("in visual mode") – that is, with the video camera in the mobile switched on.

Today, grandfather is diverting himself in the local market, which, to be specific, means he's hanging out at the popular Happy Pensioners' Club. An old friend is in the process of explaining how to use his mobile to place a

The Mobile Marketplace in 2007 – A Future Free of Surprises

bet. "It is funny," he says, "nowadays people don't say 'hello how are you?' any more. When they phone you, they say 'Where are you?' and 'Am I interrupting you?' It is really strange." Grandfather nods and tastes the beer.

"Let's make a pile of dough," they agree, raising their glasses and making a toast. Just as he has ordered another beer, his mobile notifies him that someone is trying to get hold of him.

"Oh Lord!" he exclaims, with a worried look on his face. He stares at the screen and sees that someone is ringing the doorbell at home. When the person turns round and looks into the camera, he realizes that it's his ex-wife making an unexpected visit. He ducks and wonders what the heck to do. Certainly he has no desire to interrupt this pleasant little moment at the Happy Pensioners' Club to go home and chew the fat with her. Unfortunately he has already pressed the answer button, and can thus hear a voice from the loudspeaker:

"Alan! Alan! Don't you try and hide from me! I know you're there! Who's that sitting next to you? I suppose you're guzzling beer as usual. You better let me into the house, I've got a stack of papers for you to look through …"

Almost anything can happen in the imagination, and when all is said and done, no one actually knows how the world will look in six or seven years. Much of what we have described in these stories will actually happen – we are convinced of that. Yet, more than likely we have missed other elements that will prove just as important. Whatever the case, this scenario gives an indication of how things might look in a few years' time.

In the next chapter we look more deeply into some of the insecure aspects of the mobile marketplace.

SUMMARY

- The mobile marketplace will break through with Yupplots, Moklofs and Sallies in the front rank.
- E-mail, chat and other communication services will achieve immediate success.
- Trust management and customer-led development will be central areas for marketers.
- There will be increasing focus on need segmentation.
- Sensitivity to the customer's mental state will be decisive in order to establish real communication.
- Mobile technology makes marketing and selling organizations more effective.

CHAPTER 12

Four Scenarios for a Mobile Tomorrow

What are the main uncertainties in terms of the development of the mobile marketplace over the next five to seven years? In the final analysis, as we see it, it is all a matter of human behavior. How we start using new technical possibilities, or indeed, if we start using them – these will be the deciding factors for any eventual breakthrough.

Main uncertainties

A number of uncertainties are significant for development in the future. These include:

- How will people resolve the conflict between the freedom of always being able to communicate and the curse of always being accessible?
- Will devices be sufficiently user-friendly?
- Will prices be kept at what consumers consider to be reasonable levels?
- Will consumers be prepared to divulge personal information about themselves to suppliers?
- Will consumers accept closed access portals, or will they demand open solutions from the suppliers?
- Will traditional media players manage to change or rethink their paradigms? If they do not succeed in doing so – who wins?

In these uncertainties we can identify two principal dimensions:

- *Openness* versus *integrity* in the question of the willingness of consumers to hand out personal information.
- A *broad* or *narrow* breakthrough for mobile data services.

Openness or personal integrity?

How open do we actually want to be? This is a central question. Studies (including the study undertaken within the framework of this book) indicate that as consumers we do not have an excessive amount of confidence in how suppliers use the information we hand out. Our willingness to release personal information in return for special offers or advantages is also relatively low.

But this is only when the question is directly put. In practice, few people have any objections to supermarkets or boutiques registering their purchases. Many also use store cards, in spite of constant warnings in the media about how this type of information may be abused.

There is a marked discrepancy between theory and practice. In theory we are restrictive, in practice fairly open. But the prerequisite for openness is that we do not feel that we are being used. Violations of trust can very quickly swing the pendulum towards greater confidentiality.

In the same way, clearly perceived customer benefits, such as selected and personalized offers, may encourage us to hand out more and more information about ourselves. Similarly, faced with an increasing flow of information, we may welcome profiling so that service providers can select the actual information and offers which are of relevance to us.

How will people navigate in the information structures of the mobile marketplace? Will certain patterns be created – as in the Internet sphere, where users spontaneously, associatively and continually click their way onwards through a series of web services? Research seems to indicate that development is not heading in this direction. Technical limitations and a different kind of search behavior have been observed in the little that has been seen so far of WAP[95] and GPRS. Might this be an early indication that with these technologies we range within more limited fields, search less associatively and place more emphasis on tips provided by other medias and friends?

A broad or narrow breakthrough?

How broadly will the mobile Internet or mobile data services establish themselves? How many people will take an interest in the development, and how quickly will the general public take a stake? Will it remain a niche technology reaching out to a small group of early adopters and technologists, or is the mobile marketplace set to become a mass phenomenon? Will the development exceed expectations, as with the Internet, or, indeed, with mobile phones? Or will expectations be dashed?

The speed of development is obviously significant in terms of the future. But in itself, development speed is an expression of many interacting factors such as user-friendliness, cost, content and usefulness. And these, in turn, will depend on what leading players do to get the mobile marketplace moving. To this extent "speed" is a resultant variable. Nonetheless it will have decisive impact on the evolution of the mobile marketplace.

Driving forces in different directions

Determined investors looking for fast-track development … High hopes resting on the new technology. Together, factors such as these can influence the speed of development. Low prices, excellent functionality and attractive content may also act as stimulators, or at least remove perceived obstacles. Legal questions (such as copyright issues, which limit content), uncomprehensive and non-intuitive pricing, technical problems and high start-up costs are obstacles on the way to the mass market. Other obstacles are people's natural conservatism and lack of time for sitting down and learning anything new.

However, if consumers experience any tangible benefit from releasing information about themselves – in the form of customized offers, better information, and so on – this may strengthen the movement towards greater openness. Other factors, such as a desire for experimentation with new technology, may also have stimulating effects. On the other hand, poor marketing strategies, perhaps of an intrusive nature, could easily create distrust and unwillingness among consumers to help service providers construct their electronic profiles. Legal obstacles may also prove to be a spanner in the works.

A joker in the pack

Technical panic

The introduction of new technology almost always leads to resistance and anxiety. To an even greater extent, a new mass media creates fear. New medias are often associated with a sort of moral degeneration. Anyone who recalls the hard-hitting debate in the 1980s will recall that indignant politicians and anxious parents were convinced that their children would become crazed serial killers after watching *The Texas Chainsaw Massacre*. (It goes without saying that this film would magically appear in every child's video collection.) There are countless examples in history of this kind of moral panic. Perhaps the most unpalatable of these, at least in the 20th century, was the so-called McCarthy era in the US after the Second World War, which was nothing less than a witch-hunt of supposed communists in Hollywood.

New technologies often bring about a widening of frontiers. Social norms, consisting of traditional values, undergo rapid change. Those who are not concerned about the mental and social effects of a new technology, focus their attention instead on the perceived physical dangers that may threaten the user. In the early days of rail transportation, there was a commonly held fear that the speed would cause brain damage.

Mobile phones – a health risk?

At regular intervals, research of varying scientific merit presents the view that mobile phones are damaging to human health. The truth is that as yet no research institutes are in a position to give a definitive answer as to the question of mobile phones and human health. In spite of the lack of any real evidence, there seems to be widespread anxiety. The whole issue of mobile phones and health, in this context, may be referred to as a joker. The consequences for the future of the mobile phone if harmful effects were ever proved, would obviously be significant.

What is causing anxiety?

The main cause of worry is the radiation that emanates from mobile phones. Both sub-station masts and mobile phones are therefore viewed as possible health risks. Radiation emitted by a mobile phone takes the form

Table 12.1 Examples of the effects of various radio transmitters

	W
TV transmitters	20 000
VHS transmitters	2000
GSM sub-station (rural)	20
GSM sub-station (urban)	2–10
Mobile telephone	0.0004-8

Source: Information Coning: www.coning.se/info/mobilhalsa.html

of radio waves, no different from the radio waves emitted by devices such as radios or televisions. Radio waves have various frequencies. For instance, a GSM base-station uses a different frequency from a television or radio station. The obvious reason for this is that when information is beamed out it needs to be kept separate from other information. Radio wave receivers need to be receptive to one particular frequency and not another. A comparison between GSM transmitters and television or VHS transmitters, shows that the GSM transmitters have a relatively low effect.

If radio waves collide with a water molecule, they have the characteristic of transferring their energy to the molecule in question – which is precisely what happens in a microwave oven. When using a mobile phone, what we are actually doing is pressing a small radio transmitter against our heads for shorter or longer periods. The radio waves of the mobile phone heat up the brain tissue (which consists mainly of water). Even with a marginal difference of temperature (0.1%) it remains to be seen whether this might actually be harmful. A recent, much-publicized study warned about the particular vulnerability of children, because of their thinner and smaller skulls. The radiation changes the cells in the body, causing phenomena such as insomnia, headaches and memory loss.[96]

Mobile phones transmit at various levels, depending on the strength of the signal being received from the sub-station. In order to save batteries, mobiles work at lower levels when sub-station signals are strong. However, if sub-station signals are weak, the mobile will increase its own signal. The greater the number of mobile phone users, the greater the need for sub-stations – as each is only able to process a certain number of calls. In other words, if there are more substations this will reduce the strength of mobile phone signals. If public anxiety keeps increasing, this may hold back the expansion of the mobile network. As we have already indicated, greater numbers of users will necessitate a commensurate increase in the numbers of substations. Substation radio masts are usually directed towards the horizon and thus do not expose people standing in close proximity. The strength of the signal decreases markedly over distance. Generally, it is believed that this form of radiation makes up a very small part of the total radiation to which people are exposed.

Another worry is the supposed connection with cancer or brain tumors. Scientists have begun to study mobile phone users who have cancer, but as yet no connections have been established between the diseases and radiation from mobile phones or any other sources.

In the UK there are discussions about whether all new mobile phones sold should come with a brochure outlining possible health risks. The reason for this is the scientific report alluded to above, as well as fierce criticism of the British government's handling of Creutzfeld-Jacob disease. The government has no desire to create a similar situation by underplaying any possible dangers.

The car and the mobile – a bad combination

The most common health risk associated with the use of mobile phones, however, is not caused by radiation but by traffic accidents. One survey has shown that the likelihood of an accident increases by up to four times while the driver is using a mobile phone.[97] Not even hands-free sets reduce these statistics. However, the findings give rise to other interesting questions, such as the dangers posed by other distractions (passengers, and so on) or, indeed, the desirability of receiving positioned advertising messages while driving. Researchers have discovered that the heaviest users of mobile phones have road mortality rates more than twice as high as those of light users. However, there may be extenuating circumstances. It has been noticed that mortality rates for long-term users are going down – this may be a sign of users learning to combine motoring with mobile phones more safely. At the same time it is worth pointing out that many accidents have a happier ending *because* of mobile phones – reducing the time lag between the actual accident and the arrival of the emergency services.

Tomorrow's mobile marketplace – four scenarios

We are now going to go a little deeper into the various alternatives arising out of the combinations of the insecurities mentioned above. The four-sided graph in Figure 12.1 illustrates four scenarios that seem feasible on

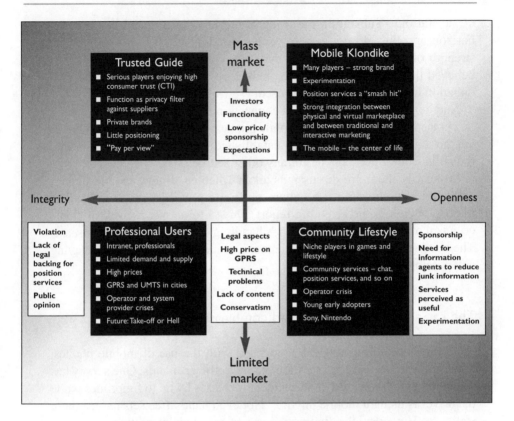

Figure 12.1 Overview of the four scenarios

the basis of the two major insecurities. We have named these scenarios Mobile Klondike 2007, Trusted Guide 2007, Professional Users 2007 and Community Lifestyle 2007.

Mobile Klondike 2007

The evolution of the mobile marketplace has exceeded all expectations. The vast majority of all mobile telephony users in the Western world, quite simply, are using mobile data services. Enthusiasm for the technology is great, there is a relatively high willingness to pay for it, and the consumers are even prepared to indulge in a bit of experimentation. The level of user-friendliness means that the sorts of search habits we're seeing are reminiscent of the behavior we associate with the Internet. Openness to experimentation in all demographic groups makes it relatively easy to introduce new services, rapidly establish a customer base, and even achieve profitability. The privacy debate that was raging at the start of the decade has fallen silent. Most people regard their personal details as a financial commodity. They are well aware of the fact that the information they provide about themselves is analyzed and sold on to others, but they consider the advantages greater than the risks.

Position marketing has become something of a killer application. Position information has been developed into a wide range of functions. Among service delivery companies and marketing companies, position commerce has become a concept. Consumers use position information to locate friends and acquaintances, set up meetings, search for retailers, and so on. And marketing companies use the application to feed consumers a continuous line of customized advertising specific to time and place. Many service companies of a similar type to McDonald's or Shell, buy position information in order to obtain a prognosis of the sorts of customer streams passing by their trading locations; and, with the help of this, are better able to modify their services in relation to "the moment of truth."

Consumers view the mobile as an extension of themselves. For instance, they use it to respond spontaneously to advertising on the television or radio. The implication of this has been an almost total integration of traditional and digital, interactive advertising. Marketing companies are able to follow the consumer further into the complexity of the purchasing decision, which is of great value once the consumer actually faces the great number of available products in the retailing outlet. Programs that help consumers find the best buys have proliferated and, to a certain extent, are a threat to retailers. Many shopping malls (together with retailers) have also developed their own local search engines and increased their customer service levels. Almost all advertisements include a bar code that enables readers, using their mobiles, to access more information or even make a purchase. The mobile is even used as an information channel in the shops. Most people use their mobile to download recipes, rather than using scraps of paper. In locations where there is no danger to traffic, traditional poster sites have been exchanged for digital screens with moving images. Many of them have even been equipped with smart technology, so that they can identify passers-by and adapt the advertising messages accordingly.

Many high street retailers are redefining their roles. They increasingly view their outlets as "windows to the world" where consumers can see, feel and touch the products; but a growing share of total sales is being generated over the net. Many retailers are practically outlets for visual demonstrations – having a last good look at products. Many products are not held in stock. The cheapest option is to place the order in the shop by pressing a few buttons on the mobile. Most often, consumers will then receive a so-called DMP (Direct Mobile Purchase) price reduction. If consumers choose to order the product later, the price might be as much as 10 percent higher. The most expensive option will be to buy products that the retailer is actually holding in stock. Discount systems and prices are a jungle in which all retailers and manufacturers seem to have their own constantly changing models.

Critics that a couple of years earlier did not believe it would be possible (in a cost-effective sense) to customize information for individual needs with time-and-place specificity, have had to withdraw all of their reservations. Recent developments in advertising have admittedly given rise to new job categories, such as MAD Editors (Mobile Advertising) – but many of the processes have been automated. Advertising specific to time and place is therefore also a feasible proposition for smaller companies. An entirely new sector has emerged to provide these kinds of services.

The mobile phone as a gadget has grown in importance, and people put a great deal of energy into finding the right kind of model that mirrors their personality and provides the right functionality. The mobile is used as a storage facility for more or less anything. It is used as an information bank, communication link, calendar and address book – in fact, anything of a personal nature. Most consumers are satisfied with one powerful mobile, to which they attach a host of external devices such as glasses with digital screens, keyboards, and so on. Developments that began to appear at the end of the 1990s, such as the Nokia Communicator or the Psion or Palm Pilot range of handheld devices, has grown into a burgeoning and many-faceted sector. Every self-respecting manufacturer of consumer electronics – and even fashion retailers – is marketing its own range. The largest market share this season is held by H&M with 6.2%, followed by Gap (5.9%) and Swatch (5.7%).

Trusted Guide 2007

Developments over the last few years have exceeded some people's expectations – but far from everyone's. The transition from a 2G to a 3G net was relatively quick. In spite of the fact that many users had upgraded to GPRS, the majority wanted to subscribe to the new networks. Curiosity, in combination with powerful financial incentives from mobile phone operators and newly established mobile services providers, meant that large numbers of people succumbed to the temptation. However, areas of low population density in the Western world are still without 3G coverage – and look set to remain so. The cost of this development has forced operators and system suppliers to dig deep into their reserves.

If anything, it is the privacy issue that has put a dampener on progress. A ferocious media debate at the beginning of the decade focused a great deal of attention on privacy. Many mobile phone users have thus chosen not to use position services, and blocked their network providers from including them in the service. A considerable number of people use the mobile as a membrane and a filter against unwanted information. Screening services have become increasingly common. Marketing and advertising organizations are putting a great deal of effort into developing more sensitive communications.

The winners in the game of earning customer approval are the large, strong, secure companies that customers are prepared to entrust with their personal details. In financial terms we are able to follow this development in the share index. Share values have in recent years been influenced by the monthly so-called CTI Reports (Consumer Trust Index), compiled by independent ratings companies such as the Consumer Trust Institute. Analysts and number crunchers believe there is a correlation coefficient of around 0.7–0.8 between CTI and the share values of these consumer companies.

The emphasis on privacy has therefore made it difficult for smaller players to establish a presence and direct link to the customer in the mobile marketplace. Content and service providers have taken on the role of sub-distributors for large portals that play the part of the gatekeepers on behalf of the stable and secure companies with high CTI ratings.

Many of the dominant portal owners are themselves mobile network operators, but there are also a number of high street retailers that have established themselves as portals. Wal-Mart is one of the biggest international players in this area. This has also led to an upswing in various private brands in the services area. Rather than bringing in various sub-distributors, many portals are choosing to develop a range of services under their own brands. Here we can see a parallel with what is happening in retailing, with the development of own-brand products.

But not all content can be categorized under the private brands bracket – the vast majority is provided by a wide range of services and content providers. A typical, large portal will feature around 10 000 content providers. Making deals with these is a science in itself. Therefore a whole new business has been established, specializing in the creation of content and services bouquets for the portals.

Professional Users 2007

Seven hard years – that is how many commentators have described developments over the latter period. But after seven hard years, the good years are coming. All hope has not yet faded about the revolutionary impact of third-generation telephony. As yet, no one even dares mention fourth-generation mobile telephony, even though there was talk at the beginning of the decade of its introduction in about 2010.

To some extent, it might be said that the last few years have been a preparation for the high hopes of what is yet to come. Professional use has well and truly been established. Not least in urban areas, 3G networks and W-LAN solutions are being utilized to extend internal company networks, thus giving employees instant access to information. With this, the accessibility of both companies and employees has substantially increased. Outside urban areas, above all 2.5 G networks are functioning with GPRS technology. A great deal of energy has been devoted to building up attractive content and practical services utilized primarily by professionals. There are many reasons for the failure of the mobile marketplace to establish itself as a mass market. The most significant of these has been the strong identification with privacy, which has been widespread in the Western world in the new century. Awareness of the consequences of leaving electronic footprints has left more and more people restrictive in their use of mobile services. This has been especially negative for position services, which have not achieved any significant success. A prevailing climate of information overload has also resulted in channels being shut down. Filtering or security services are legion. With advertising screens on the television, as well as selective newsgathering and other similar functions, it is increasingly difficult for advertisers to reach their customers. The advertising revenue model for financing the networks has therefore failed.

Prices are another reason for the poor development. Mobile operators chose to price the GPRS service relatively high so that consumers would not be unduly shocked once the UMTS networks were introduced. The effect of this was that very few people switched to GPRS at all. In spite of many subsequent price cuts, it proved difficult to entice people to abandon their existing system. The great majority failed to see the benefit of the new technology.

To avoid the perception of 3G as the fiasco of the century, there is an immediate need for a host of new services creating volume in the net, or at least generating revenues for service providers. Currently, very few 3G customers are using their mobiles as anything much more than advanced telephones. For this, no 3G network is required. A few years earlier, there was a wave of bankruptcies among mobile phone trading companies. Next in line are the mobile phone operators. These have already gone through large-scale restructuring in recent years, having first been sold off by the telecommunications operators, followed by waves of mergers. We hope to avoid a new wave of bankruptcies. But very little can be predicted about the future.

Community Lifestyle 2007

Not much is left of all the hype that surrounded the idea of the mobile marketplace at the beginning of the decade. Consumer demand for mobile services has not even remotely succeeded in filling network capacity or indeed the coffers of the network operators. While the operators' substantial subsidies have had the effect of switching urban users to 3G networks, very few are actually using the capacity of the system for anything more than some elementary information searches, receiving the odd message, or sending and receiving the occasional mail. There is no application or function that delivers the punch to either finance networks or ensure the survival of service providers. Outside the cities, 3G operators share the networks to reduce costs.

For telecommunications companies, the lack of interest in mobile data communications has become a gigantic problem. Share valuations have fallen in line with credit ratings, and many of the operators have already had to throw in the towel. Many system providers have also taken a beating, as they in many cases shared the start-up risks with the operators. In many cases there are legal processes under way against the network operators because of contractual breaches, and licenses have been revoked on account of the widespread failure to expand in accordance with contractual agreements.

However, everything is not doom and gloom. The mobile marketplace is still alive, if only as a subcultural phenomenon. Mobile data traffic is slowly growing, primarily among professional users looking to gain access to information from company intranets or various media, communicate effectively and use position-specific information. There is also a wide range of high-quality content available on mobile networks, as a result of the investments in content made by mobile operators in order to attract users to the network services.

But the most popular services are not information services, and professional men are not the predominant user group. In fact, the group that has most emphatically embraced the new possibilities comprises children and young people between the ages of 10 and 30, who primarily use the mobile as a platform for games and communication. Nintendo and Sony/Ericsson were the first to launch platforms for network-based games, which also functioned as telephones, cameras, and so on. Sony/Ericsson and Nintendo both control more than 10 percent of the market, which is more than most of the mobile manufacturers. Russia's Cybiko has another 10 percent of the market. Only Nokia is bigger, with a market share marginally in excess of 15 percent.

Various types of communities have emerged around games and other interests, functioning as communication nodes for growing numbers of people. Mobile portals, tied to these communities, offer members the opportunity of communicating, exchanging players and games, buying or swapping equipment, and so on. Position services have also become popular in these groups. These are frequently used, for instance, to locate other members in the immediate vicinity. Search profiles can be fed into the system, which will then identify like-minded individuals in the street. Secret signs of recognition are no longer the exclusive preserve of the Freemasons. Members of each and every community, with the help of digital search profiles and position services, can identify other members. With the help of so-called HI-services they can greet each other digitally as they pass each other in the street.

Conclusion

As a conclusion, let us pick out the most important building blocks for the various scenarios. In the overview in Table 12.2 you will find examples of key characteristics and factors playing decisive roles in the development of the mobile marketplace. What are the differences between the scenarios? Which are the most important driving forces? Using this as a foundation, are there any indications at the present time that seem to indicate we are moving towards any one of the scenarios?

Table 12.2 Scenario grids

Factor	Mobile Klondike	Trusted Guide	Professional Users	Community Lifestyle
Degree of openness	Open solutions predominant	High degree of closed solutions	Relatively high degree of closed solutions	Open solutions
Dominant players initially	No dominant players. A host of players compete to gain a hold in the market	Mobile operators, well-established retailers, trusted brands, and so on, take on the role of mediators between the customer and service providers	No dominant players. Content providers focused on corporate needs achieve success most readily	Community organizers
Degree of stability	Not stable in the long term. Development towards alliances between players with shared customer programmes and other loyalty rewards	Relatively stable, even in a climate of increasing openness	Either a pre-stadium to Trusted Guide Friend or collapse	Pre-stadium to Mobile Klondike or collapse
The role of portals	Narrow portals tied to service providers. Users proceed directly to the companies, without a detour via broad portals	Large, broad portals. Functions as a "matchmaker" and filter against service providers. Connects buyers to sellers, gathers service providers together; handles transactions	Narrow portals focusing on the needs of professional users. Aim to save users time when searching for information	Narrow "tribal" portals with a central role in the tribe. Service focused, but with additional services and offers from various companies. The portal plays the role of tribal chieftain
Mobile service providers – locations	National/regional	National/regional	National/regional	National and global

cont'd

Table 12.2 cont'd

Factor	Mobile Klondike	Trusted Guide	Professional Users	Community Lifestyle
Factor governing success for portals	Diversity. In the mass market, low prices and the ability to connect a physical and virtual presence	Trust	High functional quality	The ability to create shared values. Many users are subject to Metcalfe's law – utility grows in proportion to the square of the number of users
Service providers	All who believe that they have something to add to the mobile marketplace	Various players within the framework of department store. Large players create private brands	Utility services aimed at professional users	Predominantly narrow service providers aimed mainly at young people
Most important driving forces behind the development	Utility-focused services aimed at broad groups	The consumers' need for overview and security. Marketers have opportunity of extending into and adding value in the form of customer clubs and loyalty programs	Functional B2E solutions. Smart solutions	New experience-based community services. Maximized niche utility
Typical winners	No clear winners as yet. A diversity of players attempting to and succeeding in distributing new services	Telephone operators, department stores, and so on, with strong brands	Few winners	Lifestyle and entertainment companies with a young audience
The mobile marketplace financed via	Advertising, line rental, purchase of computers, commerce	Line rental and purchase of computers	Line rental and purchase of computers	Line rental, advertising and purchase of computers

Relationship between physical environment and the mobile marketplace	Very strong. The mobile marketplace used to entice customers into the physical environment, and vice versa	Relatively weak, primarily information search and peripheral services such as payment of parking meters	Relatively weak, primarily information search and peripheral services such as payment of parking meters	Relatively strong, but focused on interaction between people
Payment function	Plug-in bank with own brand	Bank owned by the portal	Plug-in bank with own brand	Plug-in bank with own brand
Strategies of main players	Service providers. Maximize synergy between physical and virtual marketplace, through location-specific CRM. Independent from mobile operators	Broad portals. A general "all-in-one" service for customers, attractive enough to make them stay	Getting a foot in the door, gaining access to large closed territories by means of strategic alliances	Niche portals: organize the clan
The mobile marketplace as an advertising channel	Very big. Strong integration between physical and mobile marketplace	Limited. Primarily pull-based advertising. Users search for information using bar codes, and so on	Limited. Primarily pull-based advertising. Users search for information using bar codes, and so on	Strong, concentrated around communities
Media landscape characterized by:	Great experimental drive. New media formats becoming widespread, with various payment models. Tailor-made information services	Traditional medias grouped around certain services, forming alliances with serious players within the mobile market	Mobile marketplace for real-time news and specialized services. Possibility of deeper knowledge, both in Internet portals and intranets of large organizations	Music and entertainment players utilize the mobile's interactive possibilities backed up by large brands looking to reach valuable global tribes
Mobile marketing paradigm	The customer's universe is reached via the mobile marketplace	"Trust management" – loyalty and confidence central to success	"Business to business and intranet marketing"	"Waiting for the young generation"
Most important target groups/mass consumer group	The broad mass	The broad mass	Professional users, in their work or privately to increase personal productivity	Teenagers, young people, lifestyle uses

cont'd

Table 12.2 cont'd

Factor	Mobile Klondike	Trusted Guide	Professional Users	Community Lifestyle
Net used for	Everything, primarily to save money	Everything	Raising personal productivity	Make dull periods into fun periods
Desire to hand out personal information	High, pragmatic approach	Relatively low, only to trusted players	Low	High in relation to the tribe, otherwise low
Users' attitudes and driving forces	Experimentation. Masses of different driving forces and motivations – everything from news to monetary savings. Interest for the new greater than fear of handing out information	Divided. Desire to get the maximum out of new possibilities, but also anxiety about confidentiality	Users are oriented towards utility and functionality. The mass market is still watching on the sidelines	Users driven by the opportunities of deepening their interests and building contact networks. The mass market is watching, asking itself, "how is this useful to me?"
Users' willingness to pay for services and information	Relatively high as long as it is something new	Relatively low	High	Relatively high
People's tendency to register at "Say no to advertising"-services, in other words MPS (mail preference service), TPS (telephone preference services), EMPS (electronic mail preference services), FPS (fax preference services)	Unusual	Relatively common	Common	Relatively unusual
Firewall services				

Most common services	Every possible service	Communication, payments, information, booking of tickets, and so on	Utility focused information, mail and so on. Connection to company intranets	Chat, information exchange, location of other tribal members in the locality
The size of the mobile marketplace	Very large	Large	Small	Relatively small
Information and consumption behavior	An Internet-type search pattern, associative, spontaneous surfing	Somewhat limited and held back. Based on trust: recommendations of friends and other media surveys	Selective, based on professional needs	Selective, focused on interest
Position information used by mobile users	Find shops, receive offers specific to time and place	Limited to find your way to information	Limited to find your way to information	Locating other tribal members
Handsets	Subsidized by operators and brand companies with their own mobile networks. Occasionally included with the purchase of other capital equipment such as cars, and so on. Mass market in existence for simple, inexpensive mobiles. Exclusive market for personalized and advanced mobiles	Subsidized by operators to drum up traffic, and by portal owners to tie down consumers	Subsidized by employers. Expensive, advanced mobiles dominate the market	Subsidized by operators and games companies. Some development from games platforms. Relatively cheap
Dominating technical system	UMTS/3G	UMTS/3G	GPRS/2.5G	GPRS/2.5G

POSTSCRIPT

Mobile Tomorrows

> It is along the borders of what has already been planted that new territories can be won. (Jan Carle)

In the turbulent world of the mobile marketplace it is impossible to predict in detail exactly what is going to happen in the future. What is possible, however, is to identify broad outlines and thus create a picture of likely or possible futures.

The key to success in rapidly changing, complex environments seems to be a combination of robustness and responsiveness – maintaining robust business concepts that work in different kinds of environments while at the same time remaining alert in thought and action, explains much of the success of companies working in a range of different sectors.[98] The road to robustness and responsiveness is a threefold one that we have dubbed "Thinking, Playing and Gardening." By emphasizing the need to understand the consequences of events in the external world and drawing broad conclusions from these, by testing strategies against imaginable scenarios, and by utilizing scenario methodology in business development, companies benefit not only from a stimulating and enjoyable internal process but also from much more robust business and marketing concepts (thinking). By forcefully experimenting into the future, by focusing not so much on adaptation as revolutionizing concepts, a lot of fun is had by all (playing). Finally, success with this process demands an open, supportive culture free of internal politics and territorial behavior (gardening).

In this book we have tried to draw the outlines of a new emerging world: the mobile marketplace. Our hope is that the reader will use the book as a springboard for continued strategic investigation in his or her organization(s) – or, to use another analogy, as an atlas for traveling into the future. For, as was once said: "This is not the end. It is not even the beginning of the end. This is the end of the beginning."

Good luck!

MATS LINDGREN, JÖRGEN JEDBRATT, ERIKA SVENSSON
Kairos Future AB

Kairos Future

Kairos Future helps companies to establish leading positions in relation to the future, both in theory and in action. We work at all levels in analyzing the future and external world – from strategic development to broad change. Our emphasis, however, is on strategic reorientation in the border territory between the future and strategic approaches to it.

Our ambition is to combine creative thinking with research that generates new insights and ideas about the future, and to integrate these findings with a practical approach in our consultancy assignments.

Our specialty is, together with our clients, to carry out rapid reorientation processes – developing long-term sustainable strategies on the basis of scenario analyses of the external world.

We also give lectures and organize conferences on the future both on our own behalf and for clients. The Kairos Academy conducts open or internal company training in business intelligence, scenario analysis and strategy development – from introductory to university levels. Read more about us at www.kairos.sc.

Contact us at:

Kairos Future
P. O. Box 804
SE-101 36 Stockholm, Sweden
Tel: +46 8 545 225 00
Fax: +46 8 545 225 01
info@kairos.se

Glossary

2G	Refers to second-generation GSM networks and mobile phones.
3G	3G is the term used to describe third-generation mobile networks. New infrastructure – everything from UMTS to new mobile devices – is being developed to offer new communication services. In theory, these services will be able to transfer data at speeds of between 384 kbps to 2 Mbps. 3G telephony will be introduced in Japan in 2001. One year later the high-speed system will be ready in Europe.
4G	Scientists are already working on fourth-generation mobile telephony. In theory, 4G will be able to run at speeds of 54 Mbps, but how this will actually work in practice is not yet known. 4G will not be a realistic proposition until 2010 at the earliest.
Bluetooth	Bluetooth is a radio frequency technology that makes communication possible between mobile and stationary units. Bluetooth transfers short-distance signals between telephones, computers and other devices. The idea of Bluetooth, in other words, is to wirelessly connect various types of devices – anything from a mobile to a PDA to an earpiece, or a control panel in the home to a washing machine, lighting, a boiler or hi-fi system.
B2B	B2B means business-to-business. Primarily, B2B companies are e-commerce companies targeting other companies as their core market. B2B commerce has proved the most profitable area of e-commerce.
B2C	B2C means business-to-commerce. Primarily, B2C companies are e-commerce companies directly targeting the consumer.
Content, content provider, content developer, content packager	"Content" refers to any kind of content in the form of a service or information available, in the context of this book, in the mobile marketplace. Content might be an application such as a game, film, music, a news service, and so on. Content

developers develop applications for sale to content providers, who provide content packagers with content that is finally made available to the end-user. All of these different players might of course be one and the same player, or at any rate have more than one role in the chain of content provision.

CRM — Customer Relationship Management in different forms looks set to become ever more important. Much energy will be devoted to developing cost-effective CRM solutions combining customized, position and time-critical information, and thus enabling customized, position and time-critical offers. Consumers, on the other hand, will become more conscious of the value of their personal details, and will make higher demands on what they receive back.

DAB — Digital Audio Broadcasting. DAB enables wireless communication up to 1 megabyte per/second (Mbps).

Early adopters — Early adopters, or early users, are those who adopt new technology before anyone else. Early adopters are usually not numerous; nonetheless they are important for the establishment of new technology and its eventual absorption into the mass market. Early adopters are often looking out for new possibilities, set out to be the first, try things that they perceive as being fun or useful, and view technology as an aid to progress and new lifestyles.

EDGE — EDGE is the acronym for Enhanced Data rates for Global Evolution. EDGE is a development from GSM which means that operators do not have to make any changes in the network structures. EDGE can transfer large amounts of data at high speeds, 384 kilobytes per/second (kbps).

ERM — Effective Resource Management (ERM) is a term used in management theory. ERM is an approach to managing human resources in the most effective possible way.

FACE — Acronym: Freshness, Accessibility, Customization and Exclusiveness. In a market bursting with choice and information flow, the receiver demands FACE information.

GPRS — GPRS means General Packet Radio Service. This technology makes it possible to transfer data at high speeds, up to 115 kbps. GPRS enables constant Internet connection, but with packet-based technology the connection is only actually used while the data is being transferred. GPRS also allows simultaneous voice and data communication, so that incoming calls can be answered while information is being downloaded. Europolitan was launched on 30 November 2000 as the first Swedish GPRS operator in its net.

GPS — Geographical Positioning System – a system for geographical positioning. By taking coordinates from a number of satellites, GPS equipment can determine position and sea level. GPS

Glossary

	enables positioning, but requires an unhindered view of the sky in order to function.
GSM	Global System for Mobile Communications. The GSM system was originally developed as a European standard for digital mobile telephony. GSM is now the most widespread mobile system in the world.
HiperLAN2	A global standard for wireless broadband LAN (Local Area Network). HiperLAN2 has a transfer speed of 54 Mbps.
HSCSD	High Speed Circuit Data, also known as High Speed. This enables higher speeds in GSM networks by utilizing more time slots. A normal call uses one time slot (=9.6 kbps). With two time slots a speed of 192 kbps is reached. Nokia 6210 is the first telephone to use HSCSD.
HTML	The Hypertext Markup Language (HTML) is composed of a set of elements that define a document and guide its display on the World Wide Web. Web browsers are programs that take the instructions (the HTML) and display your web page properly.
Instant reply	Instant reply is an immediate automated response technology that enables instant communication with the customer. This can substantially reduce response times and improve direct marketing initiatives.
IR	A technology for wireless connectivity through infrared light between, for instance, a PC and a printer.
IRL	The term "In Real Life" refers to a real, physical meeting between people. The term has emerged to distinguish between virtual and physical meetings and experiences.
Moklof	"Mobile Kids with Lots of Friends" is expected to be a major user group in the mobile marketplace. Moklofs are teenagers and young people up to the age of 30, constantly on the move and with extensive networks. They want to use their mobiles to keep in contact with friends, arrange meetings, chat, play games, listen to music, use picture telephony, send images, and so on.
MP3	MP3 is an open format for making large music files smaller without radically degrading their audio quality. MP3 is widely recognized as the most popular format for storing music on the World Wide Web and other components of the Internet. It is also called MPEG Third Layer or MPEG-1, audio layer 3.
MSP and MSO	Acronyms for Mobile Service Providers and Mobile Service Operators. These two expressions distinguish between players that offer mobile services without owning their own networks – instead, they lease their net presence from the network operators. Mobile Service Operators, on the other hand, own and run networks – examples of these include Telia or Comviq.

Glossary

NMT — Nordic Mobile Telephone Network. NMT was the world's first functioning mobile telephone system. An analogue system, it was used in Scandinavia, parts of Europe, the Baltic states, the Middle East and Asia. Two versions of NMT were produced – the original version which ran at 450 MHz and a later version at 900 MHz.

One-to-one marketing — The digital age makes it possible to communicate one-to-one. Earlier, it was mass marketing that held sway with its one-to-many format – that is, the same message to everyone. With current technology it is possible to customize communications to an infinite number of people.

PDA — Personal Digital Assistant – refers to a handheld device such as a Palm Pilot or Psion. Development is moving towards increasingly integrated handheld computers incorporating voice-led functions integrated with other functions.

Sallie — Senior Affluent Life Lover Enjoying a second spring is among the major user groups in the mobile marketplace. Sallies are senior people with grown-up children and the consumer power to influence the market. Compared with their age-mates they are very interested in all kinds of mobile services.

Showroom — A physical place used by e-commerce companies to demonstrate products to potential buyers. A showroom gets round the perennial problem of e-commerce: the need to touch things before you buy them. Like any other retailer, a showroom also functions as a marketing channel. Certain products are usually kept in stock, so that they can be taken away immediately.

SMS — SMS or Short Message Service makes it possible to receive text messages of up to 160 characters via the mobile. The messages are stored if the device is switched off or the user is located beyond network coverage. SMS has achieved success primarily among younger users, as it is cheaper to send a text message than to make a telephone call.

TDMA — Time Division Multiple Access is a digital mobile network in the USA, Latin America, New Zealand, parts of Russia and Asia.

TIMC — Total Integrated Market Communication is a marketing perspective that seeks to integrate all significant aspects of marketing in an optimized mix. While channels are proliferating it is more difficult to determine marketing approaches that are cost-effective in the longer term, and as a consequence of this, media approaches are being integrated more and more.

UMTS — Universal Mobile Telecommunications System is the name given to 3G mobile devices. UMTS is based on GSM and will offer a universal standard for global mobile communication. Because of the fact that many different technologies are

Glossary

	working together (GPRS and EDGE), many new services will be introduced while data transfer speeds increase. In theory, UMTS is projected to manage 384 kbps out of doors and 2 Mbps indoors. Skeptics are not convinced of these projections, however, and believe that speeds will be considerably lower.
Virtual communities	Virtual communities are virtual meeting points for individuals that have particular interests in common. There are communities on the net for every conceivable interest. Many companies also set up virtual communities based on their products, to increase user interest, build customer relationships and learn about possible faults or weaknesses in their products.
Virtual reality	An artificial environment which is experienced through sensory stimuli (as sights and sounds) provided by a computer and in which one's actions partially determine what happens in the environment.
WAP	Wireless Application Protocol is a solution to standardize protocol and interfaces for mobile devices. WAP is an open application supported by some 100 large providers.
Wildfire	"Oh hi! My name is Wildfire!" Wildfire is a virtual assistant, a digital secretary, that on the basis of voice recognition can answer phone calls, handle questions, learn to recognize your most frequent contacts, and so on. Read more about Wildfire at http://www.wildfire.com.
XML	The Extensible Markup Language. XML is a markup language for documents containing structured information and is the next generation of markup language.
Yupplot	"Young Urban Professional Parent with Lack of Time" is a professional, urban, well-educated parent of good income aged between 30 and 50. A typical Yupplot is a highly mobile individual, with his or her own laptop, mobile phone and possibly also PDA. His or her primary goal in life is to gain time, increase personal productivity, keep up-to-date and use time effectively. Together with Moklofs, Yupplots are a large and early user group in the mobile marketplace.

Notes

1 From *Data Smog, Surviving the Information Glut* by David Shenk, p. 77.
2 For a list of interviewees, see "Interviews" in the Sources section.
3 http://www.sims.berkeley.edu/how-much-info/summary.html.
4 World Development Indicators 2000, World Bank.
5 Interview, November 2000.
6 *Framtidens arbete och liv*, Åke E Andersson and Peter Sylwan.
7 *Forskning & Framsteg* 7/1998.
8 Kairos Future AB/FSI, November 2000.
9 InternetNews, 7 December, 2000.
10 From *The Social Life of Information* by John Seely Brown and Paul Duguid.
11 *How the Mind Works*, Steven Pinker, p. 300.
12 *Natur & Vetenskap*, issue number 9/1998.
13 Interview, November 2000.
14 *Unwinding the Clock*, Bodil Jönsson.
15 Interview, November 2000.
16 *Forbes Global*, 13 November 2000.
17 Interview, 12/10/2000.
18 *Svenska Dagbladet*, Näringsliv 14/10/2000.
19 Interview, 12/10/2000.
20 From the book *The Future of the Self: Inventing the Postmodern Person* by Walter Truett Anderson.
21 Regis McKenna, *Real Time: Preparing for the Age of the Never Satisfied Customer*.
22 Paul Levinson, *Digital McLuhan: A Guide to the Information Millennium*. All the following McLuhan quotes on pp. 32–4 are from this source.
23 Gregory Bateson: *Mind and Nature – A Necessary Unity*.
24 See, for instance, Ingvar and Sandberg: *Det Medvetna Företaget* (The Conscious Company).
25 Like a number of trios, quartets, quintets, and so on.
26 That is, if operators choose that particular payment method.
27 GSM networks have a data transfer capacity of 9.6 kbps.
28 *Svenska Dagbladet*, 11/06/2001.
29 Primarily the standard 802.11a.
30 Merrill Lynch, "Wireless Internet, Instant Reply?" (05/06/2000) The consultancy ASIS is predicting 1.8 billion bluetooth-equipped products by 2007.
31 *New Scientist*, 21/10/2000.

Notes

32 *New Scientist*, 21/10/2000.
33 According to Accenture: "The Future of Wireless: Different than You Think, Bolder than You Imagine." 2001.
34 According to Accenture: "The Future of Wireless: Different than You Think, Bolder than You Imagine." 2001.
35 One of the reasons is the dwindling price of mobile phone chips. GartnerGroup calculated a price of US$35 (Symposium Itxpo, Paris 6–9 November 2000).
36 *New Scientist*, 21/10/2000.
37 *Finanstidningen*, 26/10/2000.
38 *Finanstidningen*, 26/10/2000.
39 *Computer Sweden*, 03/05/1993.
40 Intelligent Agent from Hydraweb. "Hydraweb has equipped its load-balancing products with a new agent that will ensure better control." *Computer Sweden*, 27/09/2000.
41 "One of the underlying causes of the failure is that there has never been a good standard for classification of web page content. A web agent simply does not understand the information in a website unless it is specially programmed to understand that particular website …" *Ny Teknik*, 02/06/1999.
42 Anyone who wishes to read more about the struggle between agents and e-commerce sites, should consult the following articles which describe, among other things, the research process at SICS: "Agents will become the human face of the Net," *Datateknik* 1996: **16**: 17, "Agents will take care of your financial affairs by listening to rumours", *Computer Sweden*, 11/04/1997, as well as reports by Stefan Johansson and Staffan Hägg, see Sources. Check also the Project Alfebiite run by, among others, Cristiano Castelfranchi of the Rome Institute of Psychology: http://www.iis.ee.ic.ac.uk/alfebiite.
43 "An example of this is Answerlogic's new search engine Kuinn." *Computer Sweden*, 22/09/2000.
44 See the article, "Smarter e-commerce with intelligent agents," *Forskning & Framsteg*, 2000:7, as well as www.sics.se.
45 One of the advocates for this line of reasoning is Kevin Warwick at the University of Reading. In his book *March of the Machines* he predicts that machines will eventually question human authority.
46 *Headhunter*, no. 48/2000.
47 From the invitation to the launch of the new media channel.
48 eEurope 2002: KOM (2000) 330.
49 Thörnqvist, Gunnar (1998) *Renässans för regioner – om tekniken och den sociala kommunikationens villkor*, p. 97.
50 Interview, 28/09/2000.
51 Interview, 04/10/2000.
52 *Harvard Business Review*, Frederick F Reichheld, Phil Schefter, "E-Loyalty: Your secret weapon on the Web," July–August 2000.
53 Kevin Kelly, *New Rules for the New Economy*, p. 5.
54 Interview, 08/09/2000.
55 "Today's Young, Tomorrow's Adults," Kairos Future AB, 2001.
56 John Naisbitt, *High Tech High Touch – Technology and Our Search for Meaning*, p. 26.
57 Worst-placed are families with children, two out of three declaring that they do not have time to do anything except what is essential, according to a study by Kairos Future AB, 1998.
58 "Guidance on Work-related Stress. Spice of Life – or Kiss of Death?" EU Report no. 05–1999–00726–00–00–EN–REV–00.
59 Pepsi advert.

Notes

60 Dr. Jim Botkin. *Smart Business – How Knowledge Communities can Revolutionize your Company*, p. 26.
61 Advert in *Dagens Nyheter*, 17/11/2000.
62 Netsurvey, 9 out of 10 are using e-commerce sites as an information source, 04/09/2000.
63 *Dagens Nyheter*, 09/08/2000.
64 Kairos Future, November 2000.
65 *The Cultural Creatives: How 50 Million People Are Changing the World*, New York: Harmony Books, October, 2000 by Paul H. Ray and Sherry Ruth Anderson.
66 *Svenska Dagbladet*, 18/07/2000.
67 http://ekonomi24/Default.asp?!ID=33521.
68 *Svenska Dagbladet*, 28/09/2000.
69 *Dagens Industri*, 14/11/2000.
70 "The Future of Wireless: Different than You Think, Bolder than You Imagine", 2001.
71 Forrester Brief, "US Mobile Users Have No Time for Ads", 10/07/2000.
72 According to Durlacher Research Ltd., Mobile Commerce, 2000.
73 Young & Free, presentation by Bo Albertson, Ericsson, 12/09/2000.
74 Interview, 26/09/2000.
75 Michael Menduno, *Wired*, September 2001.
76 Jupiter Report: *Online Advertising Mix, Scandinavia*, 1999.
77 Mediabarometern 1999, Medienotiser Nordicom Sverige, Gothenburg University no.1, 2000.
78 Nicholas Negroponte, *Being Digital*.
79 Source: Integrated communications from the consumer end: Part 1, *Admap*, Feb 1996, NTG Publications Ltd.
80 Forrester Brief, US Mobile Users Have No Time for Ads, 10/07/2000.
81 Tendens Special, no. 2–4 2001.
82 *New York Times*, 29/06/2001, article: Forecasts for Spending become Even Gloomier.
83 Jupiter MMXI, research presented at the Global Online Advertising Forum in Cannes.
84 From World Bank: World Development Indicators 2000. The figure is calculated on the assumption that retailing globally is of a comparable size in relation to GNP as it is in Sweden.
85 *Computer Sweden*, December 2001.
86 *Dagens Nyheter*, "Namn och Nytt," 2000.
87 *Svenska Dagbladet*, 10/05/2001.
88 www.wvs.isr.umich.edu.
89 Accenture, 2001.
90 www.viktoria.se.
91 For a more detailed description, see *The Art of the Long View* by Peter Schwartz.
92 Frederick F Reichheld and Phil Schefter "E-loyalty: Your secret weapon on the web," *Harvard Business Review*, July–August 2000.
93 Sources: *Tendens special* (Sverige) no. 7 2000, *Marknadsfoering* (Danmark) no. 7 2001, *Adageglobal* (USA) Feb. 2001.
94 Forrester Brief, "US Mobile Users Have No Time for Ads," 10/07/2000.
95 Strand Consult: Status and perspectives of WAP technology, 2000.
96 G. J. Hyland, Physics and biology of mobile telephony, *Lancet* 2000; **365**: 1833–6.
97 Rothman, Kenneth J, Epidemiological evidence on health risks of cellular telephones, *Lancet* 2000: **365**: 1837–40.
98 60–70 percent of companies' total performance and between 20 and 40 percent of financial performance are explained by these factors. See "How much does strategic flexibility mean?" Mats Lindgren, Kairos Future AB, 2001.

Sources

Literature

Anderson, Walter Truett, *The Future of the Self: Inventing the Postmodern Person*: Putnam Publishing Group, New York, 1998.

Andersson, Åke E and Sylwan, Peter, *Framtidens arbete och liv* (The future of work and life), Natur och Kultur, Stockholm, 1997.

Bateson, Gregory, *Mind and Nature – A Necessary Unity*: Bantam Books, New York, 1980.

Botkin, Dr. James, *Smart Business – How Knowledge Communities can Revolutionize your Company*: The Free Press, New York, 1999.

Brown, John Seely and Duguid, Paul, *The Social Life of Information*: Harvard Business School Press, Boston, 2000.

Ingvar, David and Sandberg, C-G, *Det medvetna företaget* (The Conscious Company): Timbro, Stockholm, 1985.

Jönsson, Bodil, *Unwinding the Clock*: Brombergs Förlag, Stockholm, 1999.

Kelly, Kevin, *New Rules for the New Economy*: Viking/Penguin, New York, 1998.

Levinson, Paul, *Digital MacLuhan: A Guide to the Information Millennium*: Routledge, London, 1999.

Lindgren, Mats, *Scenarioplanering* (Scenarioplanning): Konsultförlaget, Uppsala, 1996.

Lundblad, Nicklas, *Teknotopier – den nya tekniken och rättens framtid* (Technotopians – New Technology and the Future of Justice): Timbro, Stockholm, 2000.

Naisbitt, John, *Megatrends – Ten New Directions for the 1990s*: Warner Books, New York, 1986.

Naisbitt, John, *High Tech High Touch – Technology and Our Search for Meaning*: Nicholas Brealey Publishing, London, 1999.

Negroponte, Nicholas, *Being Digital*: Knopf, New York, 1995.

Pinker, Steven, *How the Mind Works*: WW Norton, London, 1999.

Polak, Fred, *The Image of the Future*: Elsevier, Amsterdam, 1973.

Ray, Paul H. and Anderson, Sherry Ruth, *The Cultural Creatives: How 50 Million People Are Changing the World*: New York, Harmony Books, October, 2000.

Schwartz, Peter, *The Art of The Long View*: Currence Doubleday, New York, 1996.

Shenk, David, *Data Smog, Surviving the Information Glut*: HarperCollins, New York, 1997.

Thörnqvist, Gunnar, *Renässans för regioner – om tekniken och den sociala kommunikationens villkor* (Regional renaissance): SNS Förlag, Stockholm, 1998.

Tunström, Göran, *Berömda män som varit i Sunne*: Albert Bonniers Förlag, Stockholm, 1998.

Warwick, Kevin, *March of the Machines*: Century-Random House, London,1997.

Interviews

Avrin, Ulf, CEO, Microsoft Mobile Venture AB.

Bergquist, Magnus, Ethnologist, Viktoria Institute, Gothenburg University.

Börjesson, Martin, Technology Analyst, CRT.

Forström, Pontus, Chief Editor, *Vision* magazine.

Frankenhaeuser, Marianne, Professor Emeritus in Psychology, Stockholm University.

Gustafsson, Peter, On Position.

Hall, Peter, Telia Mobile.

Ingvar, Martin, Professor in Neurology, Karolinska Institute, Stockholm.

Leonardi, Robert, futurologist m-commerce, Logica.

Lilliesköld, Jesper and Cavenius, Håkan Intelligence AB.

Neisser, Ulric, Professor in Psychology, Cornell University, USA.

Nilsson, Alexander, Researcher, Chalmers University of Technology, Sweden.

Östlund, Britt, Ph.D Technology and Social Change. Head of Department Use and Effect of IT. The Swedish Agency for Innovative Systems.

Ström, Per, Consultant and Researcher in media.

Sturesson, Lennart, Researcher, Technology and Social Change Linköping University.

Ulfstrand, Staffan, Professor in Zoo Ecology, Uppsala University.

Wåreus, Carl, Marketing Manager, McDonald's Europe.

Wester, Hans, Hewlett Packard Sweden.

Wijkman, Lars, Research Center for Telemedicine, Luleå University.

Wikström, Solveig, Professor in Consumer Behavior, Stockholm University.

Reports

Durlacher Research Ltd, Mobile Commerce, 2000.

eEurope 2002, KOM (2000) 300, June, 2000.

Ernst & Young Financial Services, "1999 Special Report Technology in Financial Services: E-Commerce."

EU Report no. 05–1999–00726–00–00–EN–REV–00, *Guidance on Work-related Stress, Spice of Life – or Kiss of Death?"*.

Europeiska Kommissionen, "Harmonisation of turnover taxes," 1999.

Europeiska Kommissionen, "Yttrande 1/2000 om vissa aspekter på uppgiftsskydd i samband med elektronisk handel," 5007/00/SV/slutlig, 2000.

Forrester Brief, "Mobile eCommerce, Time for reality check," 27 April 2000.

Forrester Brief, "NTT DoCoMo Sets the Mobile Internet Standard," 15 February 2000.

Forrester Brief, "US Mobile Users Have No Time for Ads," 10 July 2000.

Forrester Brief, "Interactive Appliances Call for Nonintrusive Ads," 18 February 2000.

Forrester Report, "Deconstructing Media," March 2000.

Forrester Report, "The Self-serve Audio Evolution," May 2000.

Gustafsson, Peter et al., "On Position" – Mastersuppsats, EMBA-utbildningen, Handelshögskolan Göteborg, 17/05/2000.

Hägg, Staffan, "Commitment in Agent Cooperation, Applied to Agent-Based Simulation," research report 1999:11, ISSN 1103-1581.

Hewlett Packard, "Enabling new business opportunities. A Primer on e-speak," 2000.

Hyland, G.J., "Physics and biology of mobile telephony," *Lancet* 2000, **365**: 1833–6.

Johansson, Stefan, "Game Theory and Agents," dissertation. ISBN 91–628–3907–1, Karlskrona, 1999.

Jupiter Mobile Web Access, Overcoming the wireless walled gardens, July 1999.

Jupiter Online Advertising Mix, Scandinavia, June 1999.

Jupiter Strategic Planning Services, 1999.

Jupiter Trends and outlooks, Bandwith and Access strategies, 2000.

Jupiter Trends and outlooks, Web technology strategies, 2000.

KPMG, Mobile Internet, any place, any time, everything, May 2000.

Lagergren, Stridh, Queseth, Unbehaun, Wu and Zander, Telecom Scenarios 2010, research report KTH, Chalmers, PCC, and Stiftelsen för Strategisk Forskning, 2000.

McKenna, Regis, *Real Time: Preparing for the Age of the Never Satisfied Customer*: Harvard Business School Press, March 1999.

Mediebarometern 1999, "Medienotiser Nordicom Sverige," Göteborgs Universitet nr 1, 2000.

Merrill Lynch, Wireless Internet, Instant reply? 5 June 2000.

NerveWire, Independent Consulting and Intercai Mondiale, "Mapping the Future of Convergence and Spectrum Management," 2000.

Netsurvey, "9 av 10 använder e-handelsbutik som informationskälla," 04/09/2000.

Ovum Limited, Guilfoyle, Christine and Warner, Ellie, "Intelligent Agents: The New Revolution in Software," 1994.

Porter, Michael E., Sölvell, Örjan and Zander, Ivo, published by Invest in Sweden Agency (ISA), The micro-competetiveness of Wireless Valley, 2000.

Reichheld, Frederick F. and Schefter, Phil, "E-Loyalty: Your secret weapon on the web," *Harvard Business Review* July–August 2000.

Rothman, Kenneth J., "Epidemological evidence on health risks of cellular telephones," *Lancet* 2000, **365**: 1837–40.

Schultz, Don E. and Bailey Scott: "Customer/Brand Loyalty in an Interactive Marketplace," *Journal of Advertising Research*, May/June 2000.

Seipel, Peter, Stockholms Universitet, Elektroniska affärer ur ett rättsligt perspektiv (lecture), 1999.

Smith-Berndtsson, Daniel and Åström, Jonas, Handelshögskolan i Göteborg Mobila tjänster för barnfamiljer – Ett sätt för automobilindustrin att skapa långsiktiga relationer? 2000.

Strand Consult: "Status and perspectives of WAP technology," 2000.

World Development Indicators 2000, World Bank.

Magazines

Affärsvärlden
Business Week
Computer Sweden
Dagens Industri
Dagens Nyheter
Datateknik
Fast Company
Finanstidningen
Forbes Global
Forskning & Framsteg
Fortune
Harvard Business Review
Headhunter
Journal of Advertising Research
Lancet
Natur & Vetenskap
New Scientist
Ny Teknik
Red Herring
Svenska Dagbladet
Tendens special
The Economist
Wired

Index

3G 55–7, 232, 233, 234, 235, 241, 245
3G-licenses 114
4G 55

A
ABB 81
Accenture 116, 179
Accessibility 135, 136, 138, 139
ACLU 93
Affärsvärlden 164
Agent technology 74–8
Ahold 154
Amazon 5, 143, 144
Andersen, Walter Truett 27
Anoto 70
AOL 74, 159, 167
Apple 75
Archers, The 132
Arsenal 142
ASEA 81
AstraZeneca 113
AT&T 62, 163
Audi 85
Australia 59, 93, 114
Austria 114

B
Bateson, Gregory 49
BBC 128, 132
BearShare 129
Belgium 113, 114

Bell, Alexander xx, 81
Benis, Warren 110
Bergqvist, Magnus 99
Big Brother 104, 108, 142, 158
Big Champagne 129
Bikeposition 127
Biosensors 72
Birka Energi 166
Bluetooth 57, 58, 67, 220, 245
BMG 129
BMW 85, 131, 158
Boo.com 49
Botfighters 127
Bowie, David 159, 160
Branson, Richard 164
British Telecom 115, 163
Budweiser 210
BWCS 56

C
Cable & Wireless 163
Canada 93
Carle, Jan 243
Carlsson, Matts 64
Carphone Warehouse 167
Casio 169
Cerf, Vint 51
Chalmers University of Technology 64
Channel 4 142
Chelsea 220
China Electronics 85, 170
Choi, Soki 24

Cisco 87
Clinton, Bill 108
Coca-Cola 63, 86, 140
Comdex 68, 70
Compaq 169, 170
Consumer groups 118–19, 178–88
Contact-generating processes 209
Copyright issue 90
Creutzfeld-Jacob 229
CRM (Customer relationship management) 139, 151, 155, 202, 208, 214, 215–19, 237–41, 246
Cultural Creatives 110
Customer-driven marketplace 120, 122, 145
Customer-generating processes 211
Customization 121, 124, 135, 136
Cybiko 170, 236

D
DAB 55, 204, 246
Darwin, Charles 14
Data-glove 71
Data-mining 92
Denmark 114
Deutche Telekom 115, 163
Digital speech 69
Digital paper 70
Digital television 131, 139
Dismiddlesizing 83
Distribution-driven markets 120, 122, 145
DJUICE 117
DoCoMo 60–3, 73, 116
Durlacher 63

E
E2 58
Easy payment methods
Echelon 93
EDGE 54, 55, 246
EIB (European Investment Bank) 115
E-ink 70
Electric crayon 70
Electrolux 58
Elley, Peter 123

EMI 129
EMPS (electronic mail preference services 240
Endemol 159
Enemy of the state 93
Epoc 169
Ericsson 44, 49, 57, 58, 60, 66, 70, 80, 85, 112, 115, 169, 170, 171, 236, xvi
Ericsson Inside Strategy 171
ERM (Effective Resource Management) 214, 246
E-speak 77
E-street 210
EU 89
Eurosport 131
Excite 160
Exclusivity 135, 136
Exxon 210
EZ 62

F
Federal Communication Commission, US 58
Filofax 220
Financial Times 164
Finland 114
Ford 19, 64, 140
Ford, Henry 18, 20
Forrester 116, 124, 142, 158, 211
Forström, Pontus 131
Fortune 75, 164
France 113, 114
Frankenhaeuser, Marianne 16
Freshness 135
Friendposition 127
Fromm, Erich 26
FSI 177, 179

G
GAP 232
General Magic 75
Germany 59, 113, 114
Globalization 96
Glocalnet 166
Goon Show, The 132
Goyer, Matt 129

Index

GPRS 44, 45, 54–7, 170, 203, 209, 225, 230, 232, 234, 246
GPS 43, 45, 47, 70, 220, 222, 246
Grey Academy 123
Grey Global Group Nordic 123, 124
Grünewald, Per 58
GSM 54–7, 60, 88, 228, 246

H

Handelshögskolan, Gothenburg 163
Handheld computers 68
Hard Rock Café 84
Harley Davidson 184
Harvard Business Review 207
Harvey Research 210
HavenCo 129
Hellåker, Jan 64
Hennes & Mauritz 232
Hewlett-Packard 77, 169
Highbury Stadium 142
Holographic monitors 71
Homo Consumentus Effectivicus 28
Homo Consumentus Ludens 28
Homo Consumentus Politicus 28, 109
HSCSD 55
HTML 45, 46, 246
Huchison 62

I

IBM 70
ICA/Ahold 154
I-mode 48, 60–2, 88, 115
Infospace 160
Inglehart, Ronald 176, 181
Ingvar, David 200
Ingvar, Martin 27
Institute for the Future 63
Integrated marketing communication (IMC) 124, 141
Intel 171
Interactive marketplace 145
Interactive media 132, 209
Internet 3.0 129
Internet usage 88
Ipaq 170
Ireland 114

ISP 48, 60
Italy 59, 113, 114

J

Japan 59, 60, 61, 88, 114
Jobline 143
Johansson, Hans 84
Johnny Mnemonic 71
Jönsson, Bodil 16
Journal of advertising research 121, 122, 123
J-phone 62
Jupiter Communications 130, 145

K

Kairos Future 116, 176, 177, 180, 243, 244
Kari, Hannu 56
Kelly, Kevin 98
King, Stephen 159
KPN Mobile 62

L

Landau, Jon 109
Lannon, Judie 141
Lessig, Lawrence 129
Levis 144
Lewinsky, Monica 108
Lindgren, Astrid 160
London Stock Exchange 87
Longstockings, Pippi 160
Losec 113
Lucent xvi

M

Mailround.com 143
Manchester United 149
Matrix, The 71
Maybelline 210
McCarthy 227
McDonald's 30, 84, 86, 231
McDonald's Europe 125, 127, 210
McKenna, Regis 26
McLuhan, Marshall 32, 33

Index

MediaCom xxi, 134
Media consumption 132, 133
Media convergence 138
Mediatude 210
Merrill Lynch 57
Metcalfe's law 52, 238
Metro 137
Micro-geographic marketing 120, 125, 126, 127, 142, 210
Microsoft 64, 68, 69, 169
Ministry of Sound 128
Minitel 81
MIT, Massachussets Institute of Technology 71, 139
Mobile Opinion 209
Mobile Position 127
Mobile services, preferences xviii, 203, 223, 247
Mobile systems, manufacturers 168
Moore, Geoffrey 106
Moore's law 52
Motorola 133, 169, 171
MPass 127
MPS, mail preference service 240
MP3-player 45, 70
Mr Goodguy 104
MTV 129, 158
Mviva 167

N

Naisbitt, John 100, 110
Napigator 129
Napster 91, 129
Nasdaq 49
National Geographic 158
Neisser, Ullric 15
Netherlands, The 114
New Scientist 53
New York Times 8, 52
New Zealand 59, 93, 119
Nicholas Negroponte 139
Nielsen/Netratings 60
Nike 144
Nintendo xx, 230, 236
NMT 44, 248
NMT 900 61

Nokia xvi, 44, 70, 98, 115, 119, 160, 169–71, 173, 232, 236
Nokia Communicator 43, 171
NTT 61

O

OM Group 87
Omninet 205
One-to-one 124, 248
Online advertising 130
"Online Advertising Mix, Scandinavia" 130
OpenNap 129
Open Source journalism 129
Orwell, George 92
Östlund, Britt 95
Ovum 60

P

Palm OS 169
Palm Pilots 144, 170, 232
Panasonic 169
PDA 39, 45, 144, 203, 248
Peppers, Don 123, 124
Permission marketing 142, 143, 213, 215–19
Personal Chemistry 84, 206
Personalize 124
Philips 70, 74, 85, 170
Planet Feedback 109
Pocket PC 169
Pokémon 144
Polak, Fred 41
Poland 113, 114
Portals 137, 203, 210, 215–19, 224, 236, 237
Portugal 114
Position commerce 120, 127
Positioning 215–19
Post modern consumers 178–88
Post-industrial dilemma 82–3, 202
Potter, Harry 160
Product driven markets, systems 122, 145
PSA Peugeot Citroen 64
Psion 169, 170, 232

Index

Pull 211
Purdy, Gerry 110
Push-advertising 211
Push-communications 124, 211
PVR, personal video recording 74

Q
Quicktime 45

R
Rabén & Sjögren 160
Reichheld, Frederick F 97, 207
Relationship-enhancing processes 213
Retina display 71
Reuters 160
Rogers, Martha 124
Rowling, J. K. 160

S
SAAB 172
Saffo, Paul 63
Sallies 203, 223, 248
Samsung 170
Scandinavia 47, 59, 113, 119
Scandinavian Airlines Systems 86
Scenario planning 199
Schefter, Phil 207
Schultz, Don E 97
Screen Pen 170
S-curve 45–50
Seat 85
SEMA group 142
Shell 231
SICS 75
Siemens 85, 170
Singapore 59, 114, 119
SJ 81
Skanska 115
Skoda 85
Smith, Will 93
SMS 78, 142, 143, 208, 210, 213
Sonera 115
Sony, Playstation xx, 85, 129, 170, 230, 236
South Korea 59, 114

Spain 59, 114
Spectacle frames 69
Springsteen, Bruce 109
Stockholm 81
Storytelling 141
Ström, Pär 73
Svenska Dagbladet 168
Sweden 113, 114
Switzerland 113, 114
Symbian 169

T
Tablet PC 68, 69
Tachikawa, Keiji 63
TDMA 54, 248
Telefonica 115, 166
Telenor 117
Telephone operators, by revenues 164
Telia 80, 127, 160
Tesco 115
Thailand 117
Thomasson 210
Thörnqvist, Gunnar 95
Time-Warner 159
TiVo 74
Toffler, Alvin 110
Top Gear 158
Toshiba 85
Toys R Us 221
Trevithick, Richard 51
Tunström, Göran 19

U
Ulfstrand, Staffan 18
UMTS 45, 54–7, 142, 170, 203, 230, 234, 241, 248
United Kingdom 93, 114
University of Michigan 176
USA 59, 88, 93, 114

V
Victorian England 51
Viral marketing 143, 213
Virgin 156, 166
Virgin Mobile 164, 165

Vision 131
Vivendi 64
Vodafone 115, 163
Voice recognition 70
Volvo 12, 52, 63, 64, 107, 112, 149
VW Group 85

W
Walkman 44
Wall Street Journal 8
Wal-Mart 233
Walt Disney Co. 62, 86
WAP 44–7, 62, 145, 209, 217, 225, 249
Wåreus, Carl 125, 127
Warner 129
Wearables 133
Web TV 131, 145, 215
West Life 221
Wetware 71, 72
Wickström, Solveig 97

Windows CE 169
Wired 129
Wireless Car 64
W-LAN 56–7, 234
WorldCom 163
World Value Survey 176
World Wide Web 46

X
Xerox 70
XML 45

Y
Yachtposition 127
Yahoo 19, 160
Yupplots xviii, 203, 223, 249

Z
Zenith 145